Selected Issues in the Ethics of AI

Selected Issues in the Ethics of AI

Samuel Segun • Emma Ruttkamp-Bloem

Editors

Selected Issues in the Ethics of AI

Previously published in *Ethics and Information Technology*
"Special Issue: Selected Issues in the Ethics of AI"
Volume 23, issue 2, June 2021

 Springer

Editors
Samuel Segun
School for Data Science
and Computational Thinking
& The Department of Philosophy
Stellenbosch University
Stellenbosch, South Africa

Emma Ruttkamp-Bloem
Department of Philosophy
University of Pretoria
Pretoria, South Africa

Spinoff from journal: "Ethics and Information Technology" Volume 23, issue 2, June 2021

ISBN 978-3-031-19274-6

This Springer imprint is published by the registered company Springer Nature Switzerland AG
The registered company address is: Gewerbestrasse 11, 6330 Cham, Switzerland

Contents

Ethics and Information Technology (2021) 23:99–105
https://doi.org/10.1007/s10676-020-09570-y

EDITORIAL

Critically engaging the ethics of AI for a global audience

Samuel T. Segun[1] ![ORCID]

Published online: 27 November 2020
© Springer Nature B.V. 2020

Abstract

This article introduces readers to the special issue on *Selected Issues in the Ethics of Artificial Intelligence*. In this paper, I make a case for a wider outlook on the ethics of AI. So far, much of the engagements with the subject have come from Euro-American scholars with obvious influences from Western epistemic traditions. I demonstrate that socio-cultural features influence our conceptions of ethics and in this case the ethics of AI. The goal of this special issue is to entertain more diverse views, particularly those from Africa; it brings together six articles addressing pertinent issues in the ethics of AI. These articles address topics around artificial moral agency, patiency, personhood, social robotics, and the principle of explicability. These works offer unique contributions for and from an African perspective. I contend that a wider engagement with the ethics of AI is worthwhile as we anticipate a global deployment of artificial intelligence systems.

Keywords Ethics of AI · Africa · Artificial intelligence · Afro-ethics · Utilitarianism · Kantianism

Introduction

With a focus on selected issues in the ethics of AI, I contend that one of such issues, and a very pressing one at that, is the marginalisation of non-western knowledge systems in the study of AI ethics. As our world becomes more interconnected, it is evident that in every corner of world knowledge systems exist that may differ, sometimes significantly, from the predominant epistemic tradition. Unfortunately, little has been done to bridge this gap or take into consideration culture and context thereby perpetuating epistemic injustice. This injustice becomes palpable when we see discourses around digital and AI ethics skewed toward presenting Western ideals, problems, and solutions as prima facie disposition of the field. Although there is some basis to this as much of the advancements made in the field of AI have come from the West. However, with a possible global adoption of this technology, it becomes expedient to have a wider representation of ethics that accounts for diverse ethos and contexts.

Value systems differ across cultures. Birhane (2020) notes that "…Certain matters that are considered critical problems in some societies may not be considered so in other societies" (395). For this reason, an intercultural approach to the ethics of AI should inform the formation of policies and guidelines to regulate the design and use of AI. To this end, there have been recent calls for a more inclusive and intercultural look on ethics for its use in artificial intelligence systems (AIS). This is because of the growing awareness of the need for a global perspective on ethics if we intend to deploy AIS globally. One such call was made by the IEEE's Global Initiative on Ethics of Autonomous Intelligent Systems. In its call, it suggested exploring "…established ethics systems, addressing both scientific and religious approaches, including secular philosophical traditions such as utilitarianism, virtue ethics, and deontological ethics and religious and-culture-based ethical systems arising from Buddhism, Confucianism, African Ubuntu traditions, and Japanese Shinto influences toward an address of human morality in the digital age" (2017, p. 193). Another is UNESCO's (2019) decision to work on a recommendation for a global instrument on the ethics of AI, which would also serve as guidelines for practitioners, governments and policy-makers.

This issue offers a road map for carrying out scholastic intercultural dialogue in the ethics of AI. A refusal to have a broader conception of the ethics of AI would amount to what Fricker (2007) calls testimonial injustice,[1] the tendency to

---

✉ Samuel T. Segun
   ssegun@uj.ac.za

1 Department of Philosophy, University of Johannesburg,
   Johannesburg, South Africa

---

[1] With testimonial injustice, we attribute less credibility to a proposition, opinion, statement or knowledge systems on grounds of prejudice about the speaker's gender, race, ethnicity, sexuality or accent, etc. The harm caused by testimonial injustice is that it deprives the

Chapter 1 was originally published as Segun, S. T. Ethics and Information Technology (2021) 23: 99–105. https://doi.org/10.1007/s10676-020-09570-y.

attribute less credibility to an idea, proposition or account based on prejudices. By offering comparative engagements around critical issues in the ethics of AI, we get one step closer to having a more balanced and robust conception of what should constitute policies, and guidelines for the development of AI systems that are sensitive to human values.

This paper is divided into three sections. In the first, I make a justification for why we need a diverse approach to the ethics of AI, showing a significant disparity in the conception of values, ethos, and moral judgement among cultures. In the second, I offer a roadmap on how we can better engage with the ethics of AI and the development of guidelines that have far-reaching effects. In the third section, I introduce readers to the six timely articles in this special issue.

Why context matters: globally engaging the ethics of AI

Demonstrating the disparity among Western conceptions of ethics and those of other cultures such as African, South-East Asian, Middle Eastern or South American is a daunting task this paper cannot sufficiently cover; instead, I show the glaring differences in value systems, using common themes in Western, African and Chinese ethical systems. I reckon that attaching geographical labels to social, cultural and ethical features is a very problematic way of ascribing homogeneity to a group as not all peoples in that geographical space share those features or values. However, there is a way to construe geographical labels around predominant features of groups. In speaking of African moral theories, Metz[2] (2007, 2017) argues that there are recurrent salient features that can be found in many sub-Saharan cultures that are not found (in the same way) elsewhere in the world. This does not mean these features cannot be found in other cultures, it just means that they are more recurrent in Africa. The same can be said of Western, Middle Eastern and South-East Asian cultures.

In many ways, African ethics, sometimes dubbed "Afro-ethics", as opposed to Western ethics, is collectivist. A similar notion is held about Confucian ethics, a Chinese ethical system. In general, both Afro-ethical and Confucian ethical systems share similarities as collectivists systems and their normative principles, especially construal of what makes a right or wrong action, rest heavily on a collectivist disposition. Unlike Western ideals, which are predominantly built on advancing individualism, Afro-ethical and Confucian moral values share principles that advance collective progress, harmony and group cohesion.

What does all this mean for the ethics of AI? The answer is simple. It profoundly shapes the nature of contributions to the field on issues of data privacy, social robotics, conceptions of artificial moral agency, moral status and patiency, autonomous weapons systems, big data and the likes. Research shows that cultures that are uniquely individualist emphasized individual rights and autonomy while cultures that emphasized relationships and harmony were often collectivist in disposition (Hofstede 2001). This means that decision-making, negotiations and conflict management are often based on group needs to maintain group cohesion rather than individual needs or preferences. The most recent evidence to buttress this point is MIT's Moral Machine project, which significantly highlighted ethical preferences across cultural and geographical spectrums.[3]

In the moral machine experiment, the researchers gathered data from millions of respondents in about 233 countries and territories, who all made a total of 40 million ethical decisions. The paper by Awad et al. (2018) showed a substantial variation in ethical preferences across cultural lines, which, as I have argued above, correlates with geographical mappings, highlighting individualist and collectivist preferences. There were consistent disparities between collectivist ethics in Asian, Middle Eastern and African nations and individualist ethics like you would find among Euro-American nations. Respondents from collectivist cultures showed little care about saving high-profiled individuals in this experiment when compared to Euro-American respondents. This could be attributed to the disparity between individualist and collectivist cultures mentioned above. In the former, the emphasis is often placed on the value of each individual as compared to promoting group interest. So, it is very easy to understand why the value of

---

Footnote 1 (continued)

speaker or knower a level-playing field to express their ideas, which many times are embraced if another speaker expresses them but does not share the conditions that made the prejudices appear in the first place. Examples abound in history where knowledge systems of indigenous people are jettisoned only to be celebrated when appropriated by European or North American anthropologists.

[2] Metz argues that there are unique interpretations of ethics found amongst the people of sub-Saharan Africa. This does not mean that they lay exclusive claims to these types of ideas. Rather, "… it means merely that certain properties have been recurrent amongst many of those societies for a long span of time in a way they have tended not to be elsewhere around the globe" (2017, p. 62).

[3] The moral machine experiment used a trolley problem-like scenario to gather responses on a variety of ethical decisions. The overall results of the survey showed a few shared principles among respondents regardless of cultural-influenced ethical preferences, such as choosing to save many over few; however, it was clear that this was done in varying degrees across geographical mapping. For instance, among South-East Asians (Japan, China) and Middle Eastern (Saudi Arabia) respondents, and unlike Western respondents, the preference to save younger characters in the scenario were less distinct. This correlates with the notion of respect for elders found among people of collectivist cultures.

an individual would be factored into the decision-making process of a person from an individualist culture.

In individualist societies, the ties between individuals and groups are loose with everyone looking after themselves and their immediate families. On the other hand, a collectivist society ensures that from birth people are integrated into cohesive in-groups that prioritise harmony, loyalty and mutual respect (Realo 1998). From anthropological studies, Hofstede (2011) suggests that much of Western societies are considered individualist and African, Latin American and Asian cultures collectivist. Collectivist cultures like those found in Africa and South-East Asia, are built on the notion of strong ties with the family and society such that it greatly influences the framing of moral obligations and values. Western values on the other hand, and by "Western" I mean Euro-American societies, are often built on individualism which promotes different sets of values that are characteristically distinct from those in a collectivist society.

The close ties between individuals and their communities in a collectivist society make guidelines to issues like privacy differ significantly from approaches adopted in the West. Among Western cultures, great emphasis is placed on data privacy and its protection. The EU's General Data Protection Regulation enacted in 2016, which extensively covers aspects of the use, transfer and protection of individual data is a good example. In contrast, in some collectivist cultures, for example in China, the concept of privacy takes a whole new meaning. Rather than see privacy in a positive light, as a way to guarantee the protection and respect of individualism by regulating an individual's interaction with the world, privacy is often dismissed and given a negative connotation (McDougall and Hansson 2002). Tam (2018) notes that privacy is considered "si yen", which translates to mean seclusion or secrecy.

As interaction with the West continues, China's conception of privacy has evolved. Even though there are new conceptions of privacy that aligns with Western individualism, it is evident that traces of the traditional conception of privacy still surround much of the conversations on data privacy and individual freedoms. Yao-Huai notes that "…at least in the relevant discussion of many contemporary Chinese scholars, the concept of privacy is no longer been limited to the earlier, narrow sense of, but now includes all personal information (i.e., whether shameful or not) that people do not want others to know" (2005, p. 8). Taking into consideration these polarised views on privacy, it behoves international organisations and corporations seeking to extend privacy laws to consider contextual peculiarities.

Another particular theme that runs through all conceptions of collectivist ethics is relationships. In some variations of collectivist ethics, at least in Confucian (Feuchtwang 2016) and Ubuntu (Shutte 2001) ethics, promoting communal relations is considered the highest good. One conception of this form of relational ethics dictates that advancing communal relations is morally good in itself since it shows respect for others based on their capacity to be human. The major proponent of this version of relational morality, and a scholar in Afro-ethics, is Thaddeus Metz. In his seminal work "Toward an African Moral Theory", Metz (2007) insists that there is a very appealing interpretation of African ethics which takes a relational framework. Hence, "An act is right if it prizes other persons in virtue of their natural capacity to relate harmoniously; otherwise, an act is wrong, and especially insofar as it prizes discordance (2016, p. 178). From this interpretation, communal relations are good in themselves not because they help us actualise ourselves or maintain some sort of peace among members of society, even though these things are seen as appurtenances. This construal offers us the platform to ground individual rights, a vital matter in the ethics of AI.

The overall nature of a relational model of ethics renders it characteristically different from Kantian and utilitarian views is its appeal to collectivist values. The moral proclivity of an agent following a relational model is to ensure the good of others and having the community take an important place in the decision-making matrix. On the other hand, Kantian and utilitarian ethics puts the agent at the centre of all decisions, making them individualistic ethics. In addition, the expected object of contemplation for the Kantian and the utilitarian are individuals. An example comes in handy here. Suppose we are to make an ethical decision on whether we ought to lie to achieve a certain goal, say get a promotion at work. We can appeal to Kantian ethics through the categorical imperative by one, contemplating on what a universalisation of that act would mean if others had to lie to achieve their goals (Korsgaard 1996). Two, we will consider if lying in this instance is using others as a means to achieve an end (O'Neill 1975). For the utilitarian, if lying would help us maximise utility while minimising pain, then we are making the right decision to lie and get promoted. Following the relational model of Confucian or African ethics, the consideration would be if lying would affect relations with others. If we had to lie to get promoted, we are likely to be considered deceitful, dishonest, mendacious and untruthful because we have used others as a means to achieve our selfish ends. Remember that in the Metzian interpretation, we consider an act wrong if it promotes selfish ends and causes discordance. In this case, the object of our attention isn't whether the action can be universalised or whether it helps us maximise utility, instead, our focus is on how it affects communal relations.

Another good example to show how these ethical systems can be dichotomised is to consider a capacity-based argument. On a strictly capacity-based argument, these ethical theories show remarkable differences in how we ought to positively relate with others. To positively relate with others, for a Kantian, we have to appeal to their intrinsic capacity for autonomy and by extension rationality; and for the utilitarian, we must appeal to their intrinsic capacity to feel pleasure or pain. However, from both the Confucian and Afro-ethical perspective, we are enjoined to treat others rightly simply by virtue of them being humans and having the capacity to be part of a communal relationship or a recipient of such relationship (Metz 2013; Bell and Metz 2011). Even though these are all capacity-based views, the intrinsic capacities we appeal to differ; what the collectivist view affords us is a wider spectrum to accommodate other species outside humans and animals into our moral circle. Since inclusion into our moral circle is dependent on if an entity is being identified with or exhibited solidarity towards, as Metz suggests, we can consider certain artificial intelligence systems like care robots, sex robots, robot nannies as morally significant beings.

Another difference between a collectivist and individualist outlook on ethics is with the principle of autonomy. Autonomy as self-governance and in the Kantian sense of an agent having the capacity to act with reference to objective morality rather than under the influence of desires does not exist in this way in collectivist cultures as decision-making is expected to be influenced by group dynamics and needs. Each member of society is expected to be an active member willing to put aside their needs for the good of society. For the utilitarian, one becomes autonomous if they can make moral judgements that help them maximise utility. In the collectivist ethics of Confucianism and Ubuntu (a moral theory from Africa), true autonomy involves having "rich social relations" (Bell and Metz 2011, p. 83) with others. Even though these two ethical systems vary in their interpretation of how one ought to relate with others, it is clear that the community plays a central role in ethical decision-making.

Looking again at the relational model of ethics found in collectivist cultures, it is clear that these cultures prize unity, cohesion, harmony and selflessness over individuality, personal autonomy and the likes (Hofstede 1991). Also, in collectivist societies, the emphasis moves from individual rights to family and community rights as individual actions have consequences not just for the individuals but also for the community; for this reason, individuals are implored to act right for the good of all. In individualist societies, rules are often made to protect the rights of individual members of society and their freedom to express those rights. The goal of rules in collectivist societies, as opposed to individualist societies, is to foster support, cooperation, and a sense of community among members of society. A simple way to perpetuate this stands is by insisting that the needs/right of all outweighs the need of one. This conception would have an impact on how we conceive of ethics by design and ethics for design.[4]

In studying Confucian and Afro-ethics, it is easy to see certain constants that reflect features of collectivism. The same can be said of individualism when you look at utilitarian and Kantian conceptions of right or wrong actions. For Kant, autonomy is central to moral decision-making. Also, the interest of the agent is keenly factored into the decision-making process. If we take a look at Kant's recommendation in the categorical imperative, for us to visualise universalising our actions to appraise if they be right or wrong actions, the agent has to ask, "Would I want this action to become universalised given these circumstances?" The answer would suggest, at least in that context, what action should be taken. In taking this introspective journey, the agent would have to hypothesise been on the other end of the stick if that action is meted on them. For the utilitarian, right and wrong acts are constructs that can either be identified by hedonic variables, preferences, or the lack of pain. This aligns very much with an individualist way to conceiving what is right and wrong. The ability to take action on grounds of its ability to provide some sort of satisfaction is heavily individualistic. On the other hand, Afro-ethics dictates that the needs of the group should be our criterion for telling the right actions apart from wrong actions. The implication would be that individuals may be unable to maximize their satisfaction as utilitarianism enjoins us to. They would more likely have to relinquish some level of satisfaction for the good of the community.

In this section, I have shown how cultural peculiarities across geographical mappings could influence ethical preferences. It is noteworthy that we consider this when engaging in debates on the ethics of AI and recommendations for a global guideline is proposed. Also, accepting contextualised contributions is an evident way to give a global outlook to the ethics of AI. In the next section, I proposed a roadmap that will allow for global participation in discourses around the ethics of AI.

---

[4] Dignum (2018) offers an insightful way to understand the ethics of AI across three categories. First, ethics by design, which focuses on the technical capacity and integration required to develop autonomous systems that have ethical reasoning capabilities. Second, ethics in design, which addresses governing guidelines and engineering methodologies that are required to analyse the ethical implications of autonomous intelligent systems as they become more ubiquitous. And third, ethics for design, which refers to codes of conducts, guidelines, standards and certifications that guarantee the quality and veracity of engineers, developers and users who engage with research and deployment of AI systems.

Roadmap to global engagement with the ethics of AI

Having acknowledged and shown that ethics is often influenced by cultural and contextual peculiarities, I succinctly offer three recommendations that serve as roadmaps to having an appreciable engagement with the ethics of AI. I reckon that my suggestions in this section are not exhaustive. However, these counsels offer practical ways designers, engineers, developers and researchers in AI can engage ethics at a global level. More so, it addresses the three subareas of the field according to Segun's (2020) classification—robot ethics, machine ethics and computational ethics.

One way to ensure that we have global participation with the ethics of AI in such a way that we have varying voices is to insist that international bodies that seek to create guidelines and recommendations around the use and deployment of AI technology do so having had a wide range of participation from representative groups. With only a few exceptions like the UNESCO's first global standard-setting instrument on the ethics of artificial intelligence in the form of a recommendation and the IEEE's Global Initiative on Ethics of Autonomous and Intelligent Systems, very few efforts have been made to aggregate a global perspective when drafting recommendations and guidelines for the use and regulation of artificial intelligence systems. This, it can be argued, is the tendency to assume that contexts do not matter when we consider ethical issues as we expect that what constitutes right or wrong actions are universals and not particulars. I contend that one way to encourage global participation in the ethics of AI is to insist that committees and bodies that seek to develop guidelines and recommendations should take cognizance of group dynamics, context and peculiarities. In addition, drafts of such proposals should be open to contributions from scholars of varying backgrounds. This in many ways would ensure that a more robust framing of solutions is proposed.

Another viable way to encourage global participation, which inevitably would lead to having a diverse pool of experts to address ethical issues in AI, is to promote scholarship from non-Western perspectives. This special issue is one of such. Deliberate efforts need to be put to encourage an intercultural and comparative take on serious debates in the ethics of AI. The challenge to this approach has been the continued marginalisation of non-Western epistemic traditions. The continuous display of testimonial injustice from publishers, journal editors and reviewers stifle efforts to have this type of dialogue. Most times, reviewers, judging from a Western epistemic grounding, assume that alternative perspectives do not offer any worthwhile contributions. This sort of thinking is exacerbated by the fact that much of what counts as international publishing avenues (journals and publishers) are resident in Europe and America, which continue to perpetuate and expand Western ideals and influences (Chimakonam 2017). The uneven power distribution makes it difficult for wider participation on discussions around ethics, especially with respect to AI. Hence, I suggest a deliberate call to have a global conversation on the ethics of AI.

The third approach is to have funders of research into the ethics of AI support projects that offer diverse views on ethical preferences especially as they apply to artificial intelligence systems. MIT's moral machine project offers us a good look at what diverse opinions on ethical issues may look like. With more funding for projects promoting ethical contributions from marginalised epistemic traditions, we are likely to have more research and published works that showcase prima facie attractive ethical features from non-Western perspectives. Funders should approach entertaining diverse views as an opportunity to support projects that might offer alternative and sometimes better solutions to problems in the field.

By acknowledging that problems in the field are often not evenly distributed and solutions are not either, it becomes expedient to clamour for global participation in discussions around the ethics of AI. I aver that beyond culturally influenced geographical mappings of ethical systems, we must also consider socio-economic factors that may affect conceptions of the role of AI. A strong reason to take this concern rather seriously is that we are a few steps closer to having AI systems deployed across the world.

An outlook on the content of this special issue

In their article, "Applying a principle of explicability to AI research in Africa: should we do it?" Mary Carman and Benjamin Rosman emphasised the need to contextualise AI in and for Africa. The authors propose a different way to conceive of the principle of explicability such that decision-making systems that are powered by AI are fair and just and their recommendations or decisions intelligible for the context they are applied to. This means guarding against biases including racial biases in training data and building systems that are sensitive to African interests and needs.

With a complicated history with the West, Africa continues to experience imposition of ideas, values and technologies that are often not context-driven, the authors warn against uncritically adopting principles under AI for social good proposal, with one of such principle being explicability. By proposing an epistemological and ethical look at the principle of explicability when applied to core areas of machine learning, in other words answering the questions, "how does it work and who is responsible for how it works?", Mary Carman and Benjamin Rosman offer us unique ways to contextualise the purpose of an AI system to better suit a society. Having considered that Africa plays little role in the development of AI technology, developers, engineers and big tech corporations who intend to have their products used within the continent are enjoined to ensure that values with those of the society are aligned by consulting subject matter experts and knowledgeable stakeholders.

In the article, "An Ontic-ontological Theory for Ethics of Designing Social Robots: A Case of Black African Women and Humanoids", John Lamola offers us a phenomenological theory that should motivate ethical consideration by robot designers and engineers, suggesting that the socio-aesthetic state relating to black women can be used as remediation of design decisions. Lamola develops what he calls "a novel appreciation of the aesthetic and phenomenal ontology of humanlike socially-situated robots, which serves to account for their affective potentialities". He contends that the socio-political issues of race and gender play central roles in design decisions, making it crucial for designers of humanlike social robots to be conscious of their prejudices, ethnic and socio-cultural commitments while designing.

In pursuit of having socially-conscious roboticists, Lamola advocates for value sensitivity in the design of humanlike robots. Value sensitive design is a theoretically-based approach to technology design that takes into account human values throughout the process of design. Within the ethics of AI, value-sensitive design is a delicate matter requiring utmost sensitivity in the appropriation and representation of cultural and ethical values. Lamola's article calls for a wider dialogue with innovators, designers and researchers on robotics to consider the role race and gender play in the design of social and assistive robots.

Is it ever possible to consider AI systems as persons? In his article, "Artificial intelligence and African conceptions of personhood", Christopher Wareham answers this question offering a characteristically African ethical perspective. He avers, following an African account of personhood which he admits is generally anthropocentric, that we can, in fact, accommodate artificial intelligence systems as persons. The paper explores the criteria and circumstances on which personhood in a moral sense can be conferred upon AI. By leaning on the relational model of ethics found among African normative theories, Wareham notes that conceptions of personhood in Africa, unlike Western notions, are partial with a potentiality to exclude AI from our moral circle.

A relational model of ethics, which is representative of many collectivist views on ethics, suggests that the capacity to partake in communal relations would depend on how one is perceived by members of the society. Against this backdrop, Wareham proposes two ways we can conceive of AI personhood from an African perspective. First, we can think of AI as subjects of communal relations or, second, as objects of that relationship. As subjects, we would accord AI rights and duties as moral agents giving their capacity to partake in communal relations; and as objects, we would ascribe moral status to these systems as entities worthy of our moral benevolence.

In the literature, several works defend or refute the claim that AIS can be considered moral agents. One criterion that runs through much of the published works on the subject is that of computational rationality. In his article, "Computationally rational agents can be moral agents", Bongani Mabaso makes a compelling argument for why AI with computational rationality should be considered moral agents. Mabaso offers us a clinical approach to consider AI as moral agents by referring to computational rationality, an attempt at optimising decision-making process using computational resources.

Mabaso develops a logical, philosophical and computationally consistent model for building artificial moral agents and extending the debate from a purely theoretical domain to a practical one, which also accounts for artificial morality. Using a bounded-optimal computational framework to argue for the possibility of artificial moral agency, Mabaso avers that key features agency such as consciousness and autonomy are computationally possible, at least functionally. He notes that, even though complex, the framework created for computational rationality can serve as a basis to develop computational morality.

Can we have an account of moral patiency that is not typically anthropocentric? Fabio Tollon thinks this is possible in his article, "The Artificial View: Toward a non-anthropocentric account of moral patiency". In it, he provides a critique of Torrance's organic view of ethical status, a view that dictates that only moral patients can be moral agents and because AI lack sentience they cannot be the subject of moral consideration. Tollon argues that not only is this view primarily anthropocentric, it also uses this intuition to define who gets ascribed sentience and ultimately what counts as a moral patient. This would mean narrowing our moral circle and effectively alienating artificial intelligence systems who lack sentience but may even be behaviourally indistinguishable from other sentient beings.

Tollon uncovers conceptual and epistemic challenges with the organic view and advances a combination of socio-relational and an intentional stance as criteria for the consideration of moral patiency. According to this view, behavioural cues are taken into account as opposed to having maximal criteria that depend on artificial intelligence systems possessing real qualitative states. With behavioural cues, Tollon believes we can have an idea of an entity's internal state and possibly its capacity for affective states. Under this view, we ascribe patiency based on immediacy and proximate value of an artificial intelligence system to existing members of our moral circle. We consider their extrinsic capacities for social relationships and not the intrinsic capacities like sentience.

Danielle Swanepoel in her paper, "The possibility of deliberate norm-adherence in AI", questions if we should grant moral status to artificial intelligence systems. She contends that to grant AI moral status they must meet, at least minimally, deliberate norm-adherence, which they currently cannot. By deliberate norm-adherence, Swanepoel refers to acting out of reverence for the norm. Considering the need to proffer, however minimally, a framework to evaluate if we should accommodate AI into our moral circle, Swanepoel makes two critical distinctions; one, between norm-compliance and norm-endorsement and two, between deliberate norm-adherence and deliberate norm-violation.

The paper insists that we should not extend the status of moral agency to AI and if we must do so it must be with certain limits. One key reason for Swanepoel's position is that we have still not been able to answer the question of moral accountability for AI. Even if we succeed in building systems with actions that appear to adhere to norms, Swanepoel contends their inability to endorse those norms or deliberately choose to violate those norms make their adherence to the norm pretentious.

There are quite many issues to be addressed in the ethics of AI and lots of work published about them. A special issue dedicated to addressing questions around the role contexts, cultures and non-conventional ethical preferences play in the ethics of AI is a step in the right direction. I am convinced that this special issue would stir up dialogue around developing more robust and context-driven projects in the ethics of AI.

References

Awad, E., Dsouza, S., Kim, R., Schulz, J., Henrich, J., Shariff, A., et al. (2018). The moral machine experiment. *Nature, 563*(7729), 59–64. https://doi.org/10.1038/s41586-018-0637-6

Bell, D. A., & Metz, T. (2011). Confucianism and Ubuntu: Reflections on a dialogue between Chinese and African traditions. *Journal of Chinese Philosophy, 38,* 78–95.

Birhane, A. (2020). Algorithmic colonization of Africa. *SCRIPTed.* https://doi.org/10.2966/scrip.170220.389

Chimakonam, J. O. (2017). African philosophy and global epistemic injustice. *Journal of Global Ethics, 13*(2), 120–137. https://doi.org/10.1080/17449626.2017.1364660

Dignum, V. (2018). Ethics in artificial intelligence: Introduction to the special issue. *Ethics and Information Technology, 20,* 1–3. https://doi.org/10.1007/s10676-018-9450-z

Feuchtwang, S. (2016). Chinese religions. In L. Woodhead, H. Kawanami, & C. H. Partridge (Eds.), *Religions in the modern world: Traditions and transformations* (pp. 143–172). London: Routledge.

Fricker, M. (2007). *Epistemic injustice: Power and the ethics of knowing.* Oxford: Oxford University Press.

Hofstede, G. (1991). *Cultures and organizations: Software of the mind.* London: McGraw-Hill.

Hofstede, G. (2001). *Culture's consequences: Comparing values, behaviors, institutions and organizations across nations.* Thousand Oaks: Sage publications.

Hofstede, G. (2011). Dimensionalizing cultures: The Hofstede model in context. *Online Readings in Psychology and Culture, 2*(1), 2307.

Korsgaard, C. (1996). *Creating the Kingdom of Ends.* New York: Cambridge University Press.

McDougall, B. S., & Hansson, A. (2002). *Chinese concepts of privacy* (Vol. 55). Leiden: Brill Academic Pub.

Metz, T. (2007). Toward an African moral theory. *The Journal of Political Philosophy, 15*(3), 321–341.

Metz, T. (2013). The virtues of African ethics. In S. Van Hooft (Ed.), *The handbook of virtue ethics* (pp. 276–284). Durham: Acumen Publishers.

Metz, T. (2016). An African theory of social justice: Relationship as the ground of rights, resources and recognition. In C. Boisen & M. C. Murray (Eds.), *Distributive justice debates in political and social thought: Perspectives on finding a fair share* (pp. 171–190). New York: Routledge.

Metz, T. (2017). Toward an african moral theory. *Themes, issues and problems in African philosophy* (pp. 97–119). Cham: Palgrave Macmillan.

O'Neill, O. (1975). *Acting on principle.* New York: Columbia University Press.

Realo, A. (1998). Collectivism in an individualist culture: The case of Estonia. *Trames, 2*(52/47), 19–39.

Segun, S. T. (2020). From machine ethics to computational ethics. *AI & Society.* https://doi.org/10.1007/s00146-020-01010-1

Tam, L. (2018). Why privacy is an alien concept in Chinese culture. Retrieved from https://scmp.com/news/hong-kong/article/2139946/why-privacyalien-concept-chinese-culture.

The IEEE Global Initiative on Ethics of Autonomous and Intelligent Systems. (2017). Ethically aligned design: A vision for prioritizing human well-being with autonomous and intelligent systems, version 2. IEEE. Retrieved from http://standards.ieee.org/develop/indconn/ec/autonomous_systems.html.

Shutte, A. (2001). *Ubuntu: An ethic for the new South Africa.* Cape Town: Cluster Publications.

UNESCO. (2019). Elaboration of a Recommendation on the ethics of artificial intelligence. Retrieved from https://en.unesco.org/artificial-intelligence/ethics.

Yao-Huai, L. (2005). Privacy and data privacy issues in contemporary China. *Ethics and Information Technology, 7*(1), 7–15.

Publisher's Note Springer Nature remains neutral with regard to jurisdictional claims in published maps and institutional affiliations.

 Springer

Ethics and Information Technology (2021) 23:107–117
https://doi.org/10.1007/s10676-020-09534-2

ORIGINAL PAPER

Applying a principle of explicability to AI research in Africa: should we do it?

Mary Carman[1] · Benjamin Rosman[2]

Published online: 11 May 2020
© Springer Nature B.V. 2020

Abstract

Developing and implementing artificial intelligence (AI) systems in an ethical manner faces several challenges specific to the kind of technology at hand, including ensuring that decision-making systems making use of machine learning are just, fair, and intelligible, and are aligned with our human values. Given that values vary across cultures, an additional ethical challenge is to ensure that these AI systems are not developed according to some unquestioned but questionable assumption of universal norms but are in fact compatible with the societies in which they operate. This is particularly pertinent for AI research and implementation across Africa, a ground where AI systems are and will be used but also a place with a history of imposition of outside values. In this paper, we thus critically examine one proposal for ensuring that decision-making systems are just, fair, and intelligible—that we adopt a principle of explicability to generate specific recommendations—to assess whether the principle should be adopted in an African research context. We argue that a principle of explicability not only can contribute to responsible and thoughtful development of AI that is sensitive to African interests and values, but can also advance tackling some of the computational challenges in machine learning research. In this way, the motivation for ensuring that a machine learning-based system is just, fair, and intelligible is not only to meet ethical requirements, but also to make effective progress in the field itself.

Keywords Principle of explicability · Machine learning · Intelligibility · Accountability · Africa

Introduction

Developing and implementing artificial intelligence (AI) systems in an ethical manner faces several challenges specific to this technology. As research and implementation surges forward, it is necessary to develop guidelines for ensuring that we are heading in a desirable direction, ranging from assessing the moral status of AI to ensuring that the process of research and development aligns with the values we hold within our societies. For instance, at the research and development stage, we need to ask the question of how we can ensure that any automated decision-making system is just, fair, and intelligible, to

ensure that when we cede decision-making power to artificial agents, they are aligned with our values and lines of accountability can be made clear. The term 'decision-making system' is loaded and quite vague, but by this we mean systems that make use of some form of machine learning. Given that values vary across cultures, an additional ethical challenge is to ensure that these AI systems are not developed according to some unquestioned but questionable assumption of universal norms but are in fact compatible with the societies in which they operate. This is particularly pertinent for AI research and implementation across Africa, a ground where AI systems are and will increasingly be used, but also a place with a history of imposition of outside values.[1] While various frameworks and principles have been developed internationally for guiding 'Good AI', and while discussions about AI in Africa typically draw on these existing frameworks (see, for instance, Microsoft 2019), there is a notable lack of African voices contributing to these discussions. There is thus a need to critically

✉ Mary Carman
   mary.carman@wits.ac.za

   Benjamin Rosman
   benjamin.rosman1@wits.ac.za

1 Department of Philosophy, University of the Witwatersrand, Johannesburg, South Africa

2 School of Computer Science and Applied Mathematics, University of the Witwatersrand, Johannesburg, South Africa

---

1 Africa is, of course, a vast continent with many different cultures and peoples. While we talk of 'Africa' in this paper for ease of reference, we do not deny that within Africa there is great complexity.

Chapter 2 was originally published as Carman, M. & Rosman, B. Ethics and Information Technology (2021) 23: 107–117. https://doi.org/10.1007/s10676-020-09534-2.

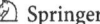

examine whether the frameworks are in fact relevant for and compatible with application in an African context.

In this paper, we take initial steps to address this need by assessing one such proposal for ensuring that decision-making systems are just, fair, and intelligible, to assess whether it should be adopted in an African research context. This is the proposal that we adopt a principle of explicability to generate specific recommendations for guiding the development of ethical AI, a principle that has not yet been assessed for African applicability. It is our contention that a principle of explicability not only can contribute to responsible and thoughtful development of AI that is sensitive to African interests and values but can also advance tackling some of the computational challenges in machine learning research. In this way, the motivation for ensuring that any machine learning-based system is just, fair, and intelligible is not only to meet ethical requirements, but also to make effective progress in the field itself. Our paper, then, firstly begins a critical assessment of the applicability in an African context of a proposal for guiding ethical AI research that has so-far been missing in the literature, and secondly builds on the motivation and support for adopting the proposal more generally. It is important to note that, while our particular focus is on a broadly-construed African context, the need for contextual and cultural sensitivity can be echoed more widely, calling attention to the need for care when drawing on generic principles that may or may not be universal in scope.

We begin by introducing what we mean by AI and machine learning, and describing some of the AI landscape in Africa. As AI research and implementation is expanding across Africa, we need guidelines to ensure that it is done in an ethical manner. So, we then turn to existing guidelines for 'Good AI' and, specifically, the European AI4People framework that identifies five guiding principles for Good AI. These are the familiar principles of respect for autonomy, beneficence, non-maleficence and justice from Western bioethics, but also the additional AI-specific principle of explicability. As the first four principles are already well-discussed within Western and African bioethics, our focus is on the new principle of explicability. We therefore engage with the question of whether the principle of explicability should indeed be adopted in an African research context, considering and rejecting two potential reasons for why it should not. Indeed, or so we argue, the relevance and importance of the principle of explicability arises from the kind of research at stake wherever it is conducted, Africa included.

Context-setting: AI and AI in Africa

Artificial intelligence is broadly taken to refer to imbuing a system with some form of computational intelligence. Under this broad umbrella, machine learning is the core technology which involves using data to optimise the parameters of a computational model, which are typically used for some form of prediction (typically regression or classification) or decision-making (reinforcement learning, over longer time horizons). The very nature of machine learning, however, as we discuss in more detail below, raises the challenge of how we can ensure that systems involving some form of machine learning are intelligible to humans, and that lines of accountability are made clear.

While machine learning and AI research and development in Africa has a long history, this has always happened in small pockets across Africa. Activity across the continent has more generally exploded over the past five years, with strong hubs forming in places such as Johannesburg and Stellenbosch in South Africa, Nairobi in Kenya, and Accra in Ghana.

In addition, several recent events and initiatives illustrate the growing interest in developing the capacity to strengthen AI and machine learning research in Africa. These include programmes such as Data Science Africa, Data Science Nigeria, and the Deep Learning Indaba. All of these aim at explicitly growing the African machine learning community, largely through technical training events and gatherings. The Deep Learning Indaba (2019), for example, is a large Africa-centric summer school which has spawned satellite 'IndabaX' events in 27 different African countries. This kind of programme has led to both growth and better organisation in the African machine learning community, as evidenced in greater participation of Africans in international conferences.

A driver for research and implementation is the vast potential for AI and related technologies to have a positive impact on communities and economies in Africa. Microsoft's (2019) white paper on the opportunities that AI offers for growth, development and democratisation in Africa, for instance, highlights four core sectors where AI could have a positive impact. These are in agriculture, by improving efficiency and effectivity; in healthcare, by improving quality and increasing access; in public services, by improving efficiency and responsiveness, and enhancing impact; and in financial services, by improving security and expanding reach. The interest in developing and implementing AI in Africa for social good does not just come from outside of the continent but can be found within Africa itself. With the hype around the so-called 'Fourth Industrial Revolution', for instance, various bodies have been set up to explore and promote the use of technologies like AI, machine learning and nanotechnology in Africa, centres such as the South African Affiliate Centre of the World Economic Forum's Centre for the Fourth Industrial Revolution (C4IR).

Given the interest in and likely growth of future research, as well as the trajectory towards implementing more AI and related technologies, there is an urgent need to ensure that any such developments and implementations are done responsibly and thoughtfully. With regard to the powerhouse of machine learning, the need for systems that are just, fair, and intelligible is a very real need if we are to guide research in Africa in the direction we want.

Guidelines for good AI and the principle of explicability

A major focus in current AI research, from both the technical and philosophical communities, is on ensuring 'Good AI': that AI is developed and implemented in an ethical and sustainable manner. For instance, at NeurIPS 2018, workshops were held on *Ethical, Social and Governance Issues in AI, Challenges and Opportunities for AI in Financial Services, Machine Learning for the Developing World (ML4D)*, and *AI for Social Good*. In the past few years, several guidelines and frameworks for ensuring Good AI for society and Good AI research have already been drawn up. These include the *Asilomar AI Principles* (2017), the crowd-sourced 'Ethically Aligned Design: A vision for prioritising human well-being with autonomous and intelligent systems' (2017), Microsoft's white paper titled 'Artificial Intelligence for Africa: An opportunity for growth, development and democratisation' (2019), and the European AI4People's publication, 'AI4People—An Ethical Framework for a Good AI Society: Opportunities, risks, principles, and recommendations' (2018), where this last framework surveys a range of guidelines and frameworks to develop a synthesis of existing principles.[2] With the proliferation of frameworks and guidelines, it makes sense to examine their commonalities and so our focus is on the AI4People framework because of the synthesis it offers of other frameworks. Through their synthesis, the authors identify five recurring ethical principles that are recognised in one way or another by all of the guidelines surveyed by the AI4People project. In this section, we introduce the five principles it identifies, the four Western bioethical principles of respect for autonomy, beneficence, non-maleficence and justice, along with a fifth principle specifically for AI, the principle of explicability, which is our focus.

The principle of respect for autonomy is roughly 'the idea that individuals have a right to make decisions for themselves about the treatment they do or do not receive' (Floridi et al. 2018, p. 697). Applied to AI where we might 'willingly cede some of our decision-making power to machines', the principle requires 'striking a balance between the decision-making power we retain for ourselves and that which we delegate to artificial agents' (p. 698). The principle of beneficence, in turn, requires 'promoting well-being, preserving dignity, and sustaining the planet'—basically, developing AI technology that benefits humanity (p. 696). The closely related principle of non-maleficence is one of doing no harm, requiring avoiding certain overuses and misuses of AI technologies. The fourth principle of justice typically requires the fair distribution of goods and services. Applied to AI, justice might require using AI to right previous wrongs, ensuring that the benefits of AI are shared fairly (and, presumably, that the burdens are fairly distributed), and ensuring that any new harms are prevented (p. 699).

The fifth principle is the principle of explicability. In a context where a select few are leading the way in the development and implementation of AI technologies that either directly or indirectly impact the rest of society, the various surveyed guidelines call for 'the need to understand and hold to account the decision-making processes of AI', while recognising that the workings of AI 'are often invisible or unintelligible to all but (at best) the most expert observers' (Floridi et al. 2018, p. 700). As the authors describe it, the principle of explicability should be understood in both an *epistemological* sense of intelligibility and in an *ethical* sense of accountability.

In the epistemological sense, the principle asks for an answer to the question of *'how does it work?'*. This epistemological sense can be found in the Asilomar AI Principles (2017), for instance, as a requirement for 'failure transparency': 'if an AI system causes harm, it should be possible to ascertain why'. The General Principles of 'Ethically Aligned Design' (IEEE 2019) call for a need for the basis of a decision to be discoverable, as does the Partnership on AI (2018), calling for 'the operation of AI systems to be understandable and interpretable by people, for purposes of explaining the technology'. The need for transparency and explainability is identified in the European Group on Ethics in Science and New Technologies (2018) and in the UK House of Lords Artificial Intelligence Committee report (2018). All of these call for an answer to the epistemological sense of *'how does it work?'*.

In the ethical sense, the principle of explicability tackles issues of accountability by asking for an answer to the

---

[2] The guidelines and frameworks surveyed are: The *Asilomar AI Principles* (2017), the *Montreal Declaration for Responsible AI* (2017), the *General Principles* in the IEEE Global Initiative's 'Ethically Aligned Design' (2017), the *Ethical Principles* in the 'Statement on Artificial Intelligence, Robotics and "Autonomous" Systems' of the European Group on Ethics in Science and New Technologies (2018), the principles of the 'AI in the UK: Ready, willing and able?' report of the UK House of Lords Artificial Intelligence Committee (2018), and the *Tenets* of the Partnership on AI (2018).

question of *'who is responsible for the way it works?'*[3] For instance, the Asilomar Principles include a requirement that the designers and builders of AI systems have a duty to shape the moral implications of the use, misuse and abuse of those AI systems. Both the Partnership on AI and the European Group recognise that AI research and development needs to be accountable to a range of stakeholders, with the House of Lords report calling for clear lines of accountability.

The principle of explicability is valuable on a number of fronts. Firstly, it addresses the uneven power structure already apparent in the development of AI, between those who are developing the technologies (typically large corporations) and those who will be affected by them (the consumers and the rest of society). Secondly, it both complements and enables the other four principles. For AI to be both beneficent and non-maleficent, we need to understand what kinds of benefits and harms it can actually do within a society. Similarly, if we are to respect human autonomy, we need to know how an AI system would choose and act. Additionally, for the principle of justice to be respected, we need to ensure that there is accountability. Thirdly, the principle recognises the role that intelligibility and accountability can play in engendering public trust and understanding, necessary for ensuring that AI is accepted within society. Without public trust and understanding, the potential economic and societal benefits of AI and related technologies could fail to materialise (Winfield and Jirotka 2018). For instance, a lack of trust can be seen to underlie fears about automation negatively impacting human employment. With powerful bodies like South Africa's Congress of South African Trade Unions (COSATU) not wholeheartedly behind such technologies— 'You can't be talking about the future of work when you describe displacement and unemployment' (Steyn 2018)— the benefits will be difficult, if not impossible, to achieve. The motivation that the AI4People framework draws on for the principle of explicability, which includes requirements for intelligibility and accountability, is in large part based on societal benefits such as public trust and understanding.

Towards ethical AI in and for Africa

The AI4People framework uses the five identified principles to generate a set of recommendations for Good AI within a European context, acknowledging that recommendations based on the principles may differ in different cultural contexts—at least to the extent that 'different cultural

frameworks inform attitudes to new technology' (Floridi et al. 2018, p. 701). So, if we are to apply the principle in an African research context and to the design of systems to be implemented in Africa, we need to develop our own recommendations based on principles that are culturally and socially sensitive. However, truly acknowledging the impact of different cultural contexts is not limited to appreciating that different cultural frameworks inform attitudes to new technology. It more fundamentally requires ensuring that the principles themselves are applicable.

Take, for instance, the communitarian nature of many African cultures and worldviews. While not all African cultures and worldviews are communitarian, while those that are need not be identical to one another, and while communitarian cultures and worldviews exist outside of Africa, the centrality of community is widely agreed to be a salient and dominant feature that can be found in various forms across the continent below the Sahara, and a feature that has been drawn on by those working within sub-Saharan African philosophy and ethics.[4] In many communitarian societies, people often engage in joint decision-making or refer to authority figures for guidance as part of their decision-making, thereby legitimately including others in a normal process. This is in stark contrast to a typical Western worldview that centralises the individual, and which is reflected in bioethical principles like the principle of respect for autonomy, frequently understood as respecting the decisional autonomy of an individual who makes decisions without undue coercion (see Beauchamp and Childress 2012). Similarly, the AI4People report expresses the principle of respect for autonomy in a way that highlights the focus on the individual: 'individuals have a right to make decisions for themselves' (Floridi et al. 2018, p. 697).

The salience of community versus a strong individualism illustrates why we require, on one hand, sensitivity in how we adopt and adapt the principles in different contexts, if we are to apply them. For instance, the World Health Organisation (WHO) has made provisions to allow partner agreement in reproductive research in certain countries with a cultural tradition of involving partners or family in decision-making, despite usually taking the involvement of a partner as a violation of participant autonomy (Moodley 2007, see also WHO 2020). The principle of respect for autonomy is still applied, but it is adapted to reflect the real communitarian-infused decision-making processes that people engage in.

On the other hand, we might reject that any of these principles should be applied at all. Over the past few decades,

---

[3] This question is an ethical question. There are, of course, questions about legal accountability and responsibility but we do not attend to them in this paper.

[4] For a sample of seminal philosophical work highlighting community within different cultural contexts, see Mbiti (1990) (Kenya, although with a systematic review of other cultures), Gyekye (1987) (Akan, Ghana), Gbadegesin (1991) (Yoruba, Nigeria) and Ramose (2005) (South Africa).

an increasing amount of work has been done developing the field of African bioethics in response to exactly this kind of challenge (as samples, see Murove 2005; Behrens 2013, Chukwuneke et al. 2014; Rakotsoane and Van Niekerk 2017; Barugahare 2018).

Reasons given for rejecting—or at least seriously critiquing the applicability of—Western principles include both the pragmatic and the theoretic. Pragmatically, simply adopting foreign principles that are divorced from the ethical worldviews that govern ordinary peoples' lives can result in practices that are inefficient in achieving their aims. In general, people are more inclined to accept ethical ideas or interventions if they are consistent with their own worldviews (Behrens 2013). The former director of Médecins Sans Frontiers, Roy Brauman, gives the example of emergency food supply in famine-stricken Uganda. Medical workers prioritised giving food to the most vulnerable, women and children, only to discover that the food was being taken away and given to local elders in lines with local customs that prioritise respecting social orders (Hellsten 2006, p. 73). Such an example illustrates how applying a principle like justice without sensitivity to local context can be ineffective in achieving its aims. This could be addressed by applying the principle in a culturally sensitive manner; however, and more crucially, the example also illustrates how a reliance on predetermined principles can exclude other principles that in fact govern people's behaviour.

While pragmatic issues could potentially be addressed by adapting the principles in a context-sensitive manner while also being open to the existence of other principles, a deeper theoretical set of issues remain. These are particularly pertinent in the postcolonial African context where there is a tradition of postcolonial critique and an expressed need for the reclamation of human dignity, authenticity, and a positive assertion of African identity following centuries of subjugation by Western powers.[5] As Andoh writes, drawing on this tradition, an attitude of 'assimilating Western values and ideologies into Africa can give rise to a situation of self-dehumanisation and outright self-subversion both in terms of dignity and self-esteem' (Andoh 2011, p. 69). For instance, given the salience of community in many sub-Saharan African cultures, a prioritising of individualism can sever the person from her 'relational spheres of existence' (Murove 2005, p. 27). Rather than simply adopting an individualistic principle like respect for autonomy, we might thus instead adopt a broader principle of respect for persons, which captures the essence of the principle but allows more cultural nuance (Behrens 2013). Alternatively, we could introduce

new principles like 'human life invaluableness' (Rakotsoane and Van Niekerk 2017), or even ones that capture the respect for social order brought out in the Ugandan example.

These critiques of the traditional bioethical principles put pressure on the applicability of the very same principles found in the AI4People framework for use within an African context, or indeed any other cultural context that does not share similar features to the central European context in which the AI4People framework was derived. We cannot simply, and without critical engagement, adopt the principles and use them to generate context-specific recommendations, at risk of resulting in ineffective processes or causing genuine moral harm. We first have to assess if the principles themselves can be recruited within the context at hand. Applying the principles of respect for autonomy, beneficence, non-maleficence and justice in an African AI context requires more examination, but our current focus is on the *new* principle of explicability. This is because there is already a rich literature within African bioethics, as well as the field of global bioethics more widely, examining the supposedly universal applicability of the standard principles of Western bioethics, whereas the principle of explicability is AI-specific and has yet to receive critical attention. So, while we could question what *form* such a principle would take in a particular cultural context—if explicability is closely related to communication and public trust as described at the end of the previous section, for instance, then different cultural norms of communication may inform what needs to be made explicable, to whom, when, and to what degree—we must first question whether the principle of explicability is relevant and applicable at all. This second level of critique, which is our focus, is crucial given the theoretical issues that warn against the uncritical assimilation of Western values into African contexts.

In the rest of this section, we critically assess whether the principle of explicability should be applied in an African context, rather than just what form it should take if it were applied. We first argue that the importance of the principle arises in part from the very nature of the technical research at hand, and in the following subsection we consider two possible reasons not to adopt it in an African research context, arguing that neither takes hold.

The importance of the principle of explicability

The principle of explicability, *prima facie* at least, is not obviously based on strongly Western values like the individualism that underlies the principle of respect for autonomy. In fact, it is a principle that could allow us to be sensitive to cultural nuances as a matter of necessity and, as we will suggest, a principle that arises out of the nature of much of the research in question. The epistemological sense of the principle, at least, is required to address some of the

---

[5] This is a tradition that draws on a diverse range of theorists from across the continent, such as Senghor (1988), Mbembe (2001), Wa'Thionga (1986) and Biko (2002).

computational challenges within machine learning. In this subsection, we thus illustrate how the principle of explicability ties in with some real issues and risks that computational and technical researchers are addressing, including those working in Africa, as a way of motivating for the adoption of the principle within an African research context.

The epistemological sense underpinning the principle of explicability seeks an answer to the question of 'how does it work?' There are two primary issues in which the epistemological sense of explicability is worth considering.

The first issue is that the fundamental modality of machine learning comes down to a human specifying a 'goal', or more technically an objective function, and the learning procedure is required to optimise a model, or technically the parameters of the model, for achieving this goal. This stands in stark contrast to more traditional programming paradigms where a human specifies the full set of steps (algorithm) that is to be executed by some software. Machine learning may instead, for example, require that a set of positive and negative outputs are provided to the learning system, which then infers the procedure for distinguishing between these categories itself. While this shift in attitude to problem solving has been transformative, there are at least two broad risks that arise.

The first risk is that of the human misspecifying the desired objective function. This may happen for example in reinforcement learning (RL), where an artificial agent is required to learn to take a sequence of decisions to achieve some long-term goal. In RL, desired states of the world are typically annotated with some positive reward, and undesirable states with some negative reward. In general, these attributions are arbitrary (both in location and magnitude), and it is trivially easy for an incorrect scaling or attribution to lead to undesired behaviours. A human, however, may not be fully cognisant of her own objectives, and as such may incorrectly imbue them into a learning system. An oft-cited example of this is the thought experiment of the paperclip maximiser: an intelligent agent tasked with running a facility to produce a maximal number of paperclips could self-improve to the extent that the result is 'a superintelligence whose top goal is the manufacturing of paperclips, with the consequence that it starts transforming first all of earth and then increasing portions of space into paperclip manufacturing facilities' (Bostrom 2003). The scenario shows how even this simple seemingly benign goal could be sufficient to generate behaviours antithetical to human life and flourishing.

This first risk speaks directly to what is known as the value alignment problem. The value alignment problem refers to the challenge of ensuring that the goals of an artificially intelligent system do not contradict (typically inadvertently) the values of humans, or society in general. In an epistemological sense, to ensure that the values are aligned requires seeking an answer to the question: 'how does it work?' This in itself is a nontrivial question to answer, as discussed in more detail below. In a case where the goals do contradict and go against human values, we are faced with the ethical question: 'who is responsible for how it works?' For instance, who is best able to explicitly identify all the relevant values that are often implicit within ourselves and our societies, and can we hold someone accountable for failing to ensure that the goals of an AI system cohere with some set of implicit values of humans? Would explicitly identifying relevant values using quantitative methods even be sufficient for capturing the complexity and flux of human values, something that Sloane and Moss (2019) question?

This problem is exacerbated by the fact that human values differ across different societies and contexts. As a simple example, consider the 'rules of the road' which change between countries, from the side of the road on which people drive, to the rules for entering roundabouts. This is even more notable in how societal preferences change between countries in versions of the 'trolley problem', where people around the world have been surveyed on which of two random sets of people should be spared in a vehicle collision (Awad et al. 2018). One can note, for example, that there appears to be a preference for sparing the lawful and pedestrians in Japan, versus a preference for preserving humans over animals as well as high status individuals in Nigeria.[6]

The second risk is that the general form of a machine learning model may be difficult to interpret. The quintessential example of this is artificial neural networks, which are widely acknowledged to operate as 'black boxes'. This arises from the basic modelling assumptions, that the relationships between various inputs and the desired outputs are typically learned to be complex nonlinear functions, which are recursively embedded in other nonlinear functions. As the numbers of parameters that are learned in these models can easily be in the millions, interpreting these functions quickly loses feasibility. The question of 'how does it work?' thus becomes of paramount importance, with limitations for just how much detail we can give in answer. The question of 'who is responsible for how it works?' also highlights how very few humans, if any, are able to understand the processes that are followed, yet that select few still have considerably deeper understanding than the many who might be impacted by the technology.

It is interesting to note that other classes of machine learning models do not necessarily possess this same characteristic, although they seldomly achieve the state-of-the-art performance that neural networks do. An example here is that of decision trees, which involve learning an ordering of features to treat as conditional rules for classifying data points. By following this sequence which is learned, one can

---

[6] https://moralmachineresults.scalablecoop.org/.

easily trace out and validate decisions made by these models. Different models may thus generate different requirements in terms of answering the question 'how does it work?'.

Another issue concerns biases that may be presented to the learning system in its training data. Examples of this have been widely seen in supervised learning, with numerous cases of racial biases being reported globally (Buolamwini and Gebru 2018). This is particularly troublesome when confounded with the previous challenge of difficult-to-interpret models, in that it could easily become unclear that these biases have been introduced to the system. Establishing how a system is making decisions, especially when it is fed with data that we may not realise to be ethically suspect, is therefore important if we are not simply to recreate our own human inefficiencies as decision-makers and agents. When these biases reflect real-world human biases and have the potential for profound and detrimental real-world impact, we again face the ethical question of who is responsible and who should be held accountable.

The above examples illustrate some general computational issues within current research the world over, and they are issues wherever research is taking place and systems implemented, including in Africa. What becomes apparent from these examples is that we need something like the epistemological sense of the principle of explicability not just for engendering public trust and understanding and for ensuring alignment with societal values, but for actually informing research to tackle some of the computational challenges that are being faced. This is because those computational challenges themselves require alignment with societal values and needs, in turn requiring that certain values and objectives are made explicit. We thus need to be sensitive to the values and needs of the societies where AI technologies are both developed and implemented. In the AI4People report, the authors describe the 'dual advantages' of an ethical approach to AI, which allows identifying and leveraging 'new opportunities that are socially acceptable or preferable' and 'enables organisations to anticipate and avoid or at least minimise costly mistakes' arising from 'courses of action that turn out to be socially unacceptable and hence rejected' (Floridi et al. 2018, p. 694). As we have argued, the advantages of an ethical approach go even a degree deeper than the authors originally discuss. Various challenges facing technical research are in fact 'socio-technical' in nature (Crawford 2017). As such, applying the principle of explicability, especially in its epistemological 'how does it work?' sense, does not only have the dual advantages identified in the AI4People report, but an additional advantage in that it can help solve some of the computational problems facing AI researchers, avoiding courses of action that are ineffective in addition to but distinct from prioritising their social acceptability.

Are there reasons not to apply the principle in Africa?

While the principle of explicability is an important principle for guiding research both to achieve computational ends and to strive for societal benefits, there may nevertheless be reasons not to adopt and apply it in African research contexts. Here, we consider two such reasons and argue that they do not show that the principle itself is problematic or irrelevant. Rather, the potential problems highlight at least two lessons: the principle of explicability absolutely requires contextual sensitivity in its application, and it must be balanced with other relevant principles.

A first potential problem, the trade-off problem, relates to the epistemological sense of the principle of explicability. This is the problem that we might face undesirable trade-offs in demanding explicability. One of the recommendations put forward by AI4People is that, in a European context at least, a framework that enhances the explicability of AI systems that make socially significant decisions is developed, where 'central to this framework is the ability for individuals to obtain factual, direct, and clear explanation of the decision-making process' (Floridi et al. 2018, p. 702). One way to meet this demand is by requiring that systems produce explanations of their own behaviour (see, for instance, Selbst and Powles 2017; Doshi-Velez and Kim 2017; Winfield and Jirotka 2018). Yet, requiring explanations in this manner for a system to meet explicability requirements could hypothetically mean that the capabilities of that system are severely handicapped (Wachter et al. 2017). In a similar fashion, London (2019) has recently argued that a demand for explicability or for making something interpretable is typically a demand for an explanation of causal relations. Domains such as AI and even medical decision-making, however, typically involve associations that are not necessarily causal. As London argues, in a domain where we lack causal knowledge but where predictive and diagnostic accuracy are nevertheless important, a demand for explicability can needlessly detract from accuracy and reliability.

So, if a system would lose accuracy or reliability in diagnosing some life-threatening disease by some percentage, how should we view this trade-off? This is a challenge that all societies need to address, but it is particularly pertinent in many African contexts where other solutions, such as medical professionals and state-of-the-art laboratories, are not easily at hand. As a representative example, consider Tanzania: as high as 71% of the population lives in rural, difficult-to-access areas with poor infrastructure, a fact that informs the Tanzanian government's current five-year health sector strategic plan for increasing access to healthcare services (United Republic of Tanzania 2015) and a fact that explains the welcoming of the use of drones to provide basic medical supplies (Landhuis 2017). This challenge is significant in the health sector, as a result of a number of factors such as different disease profiles around the world. Malaria, for example, poses a much greater risk in Africa than it does in Europe, and this coupled with a shortage of experts necessitates automated solutions (Brown et al. 2019).

Challenges also exist in the social sphere. Africa is home to an estimated 2,000 languages, and addressing communication barriers is an important step towards advancing these societies. The sheer scale of this problem again calls for AI-based solutions in automated translation (Abbott and Martinus 2019).

Part of the attraction of the development and implementation of AI solutions in Africa is that doing so can address challenges like these and others faced by African societies that arise from social, historical and geographical inequities that make solutions available elsewhere in the world untenable. This is something that the Microsoft white paper discussed earlier does indeed highlight, by focusing on the way AI could be used to improve various sectors, such as agriculture, healthcare, public services and finance (Microsoft 2019). In this context, the stakes for demanding explicability at the expense of accuracy or reliability can be particularly high.

This problem, however, does not show that the principle is itself problematic or irrelevant in an African context. For one thing, there are two senses of explicability contained within the principle, the epistemological and the ethical. In the epistemological sense, we might seek alternative and less demanding ways to account for how a system works. We have argued, for instance, that solving some of the computational problems facing machine learning requires making objectives and goals explicit. As such, we could plausibly achieve explicability in the epistemological sense by specifying a system's design goals more carefully. Indeed, Kroll (2018) has proposed such an approach for intelligibility, one that shifts from a focus on understanding technical tools to understanding the overall system, which includes people. Such a tactic need not require an explanation in terms of causal relations, as per London's (2019) worry. In fact, making an overall system intelligible, including the people that are part of it, may require non-causal explanations, such as functional or hermeneutical explanations and those found more widely in the social sciences. In this way, the machine learning-based decision-making system can be made intelligible, in terms of its goals and objectives, and accountability can be demanded of the entire system, which includes the people specifying those goals and objectives.

Alternatively, perhaps in a situation like the trade-off described above, we should step back from the epistemological sense and focus instead on the ethical sense of explicability, establishing a clear line of accountability, such as holding those who are specifying the goals and objectives accountable. But more generally, the principle is not intended as a standalone principle. It would still need to be balanced with other acceptable principles, such as the principle of beneficence—how can we best capitalise on AI technologies to ensure the well-being of people?—or, even, the principle of justice, concerning the fair distribution of goods or what constitutes fair compromises to ensure that, say, access to health services is available to all. This allows variability in what is demanded of a particular system with particular goals and within a particular context.

A second problem, a problem of compromise, focuses more on the ethical sense of the principle of explicability, and the concomitant demand for accountability. This is the problem that a demand for accountability could plausibly limit progress in a field where African nations and research institutions could be firmly entrenched among world leaders. The development of AI and related technologies promises to tackle African-specific problems that can aid in social and economic development, can create jobs, and is an arena where African researchers are already increasingly active. Indeed, the growth and appetite for events such as those of the Deep Learning Indaba, Data Science Africa and Data Science Nigeria shows that there is interest in upskilling in this direction. Imposing lines of accountability could result in onerous regulatory constraints on an industry we want to encourage, with parties becoming less willing to pursue potentially fruitful but risky research or implementation.

This problem is speculative and overlooks that the principle of explicability does not state what the accountability requirements are, just that we establish lines of accountability. Like the previous problem, this allows contextual sensitivity in devising recommendations from the principle, where that contextual sensitivity may consider different cultural norms regarding what needs to be explained, to whom, when and to what degree, but also must consider the cost–benefit ratio of potential regulations. Further, and as discussed with the previous problem, devising recommendations from the principle of explicability would work in tandem with other principles. For instance, justice might require differential treatment for how research is conducted in Africa, to target economic and historical imbalances between African nations and centres in the developed world. The principle itself is not obviously problematic but we have to take care with how it is applied.

Further, if Africa is to capitalise on the progress it is making with growing AI research across the continent, we do still need to ensure that it does so in a way compatible with the values and needs of those living there. In order to actually do the computational research, we will often have to address requirements that are closely related to those put forward by the principle of explicability, such as by tackling the value alignment problem. Tackling the value alignment problem requires that we identify our own objectives; that is, we must go some way towards answering the question 'how does it work?' in order to make it, the system, work for what it is designed for. If this is the case, the research and technology that we want to flourish and advance will often depend on addressing questions that the principle of explicability raises, such as demanding intelligibility. In also requiring accountability for the systems, we would be requiring accountability for something that would have to be done, at least to some extent, in order to advance the research itself.

These two problems, as we have seen, can be dealt with by highlighting that the principle of explicability would, ideally, be applied in tandem with other principles and applying it requires contextual sensitivity, not just to ensure that values are aligned with a society's values, but also to ensure that computational challenges are themselves addressed. These problems do not show that the principle of explicability itself is problematic or irrelevant.

Closing thoughts on who is accountable for how a decision-making system works

We have proposed that the principle of explicability, when applied in the epistemological sense to typical areas of machine learning research, requires identifying objectives and goals for a system that cohere with those of a given society in which the system will operate. But what implications does this have for the ethical sense of the principle and the question of who, exactly, should be held accountable for how such a decision-making system works? A third potential problem could arise here, to do with demandingness: the principle of explicability as we have developed it may be too demanding on researchers in a developing field in Africa who are frequently dependent on international input. Holding those researchers to account would be unfair. To address this problem, we tentatively propose that the demands of explicability require a division of labour, and as a result accountability could in fact be diffuse.

Suppose that the machine learning researchers based in Africa and developing a system to be implemented in an African context are to be held accountable for how the system works. As part of the demand for explicability, in the epistemological sense, we have argued that objectives and goals of the system need to be identified. But who should identify these objectives and goals?

Identifying the goals, objectives and underlying values of a society is not a straightforward matter and not something one can simply consult a rulebook for. In South Africa, for instance, it is officially required that vehicles yield right of way to pedestrians crossing at a pedestrian crossing (Department of Transport 2012). In practice, however, this seldom happens and stopping at a pedestrian crossing can surprise other vehicles on the road. Simply consulting the rulebook

would not prepare anyone, person or machine, for actual driving.

The machine learning researchers, however, are technical experts, not necessarily experts in identifying the goals, objectives and underlying values of a society with which their system's goals and objectives need to be aligned. Further, they may be contributing to global work or be part of an international research team, such as by working at one of IBM's research labs in South Africa or Kenya, or Google's research lab in Ghana, or be funded by international bodies like Google and Facebook, who fund students pursuing the African Masters in Machine Intelligence in Kigali, Rwanda. Yet, demanding that they make goals and values explicit and then holding the African-based researchers accountable for a system that is not entirely in their hands would be placing an onerous and unfair burden on them.

Identifying the goals, objectives and underlying values of a society, as those working on the ethical design of technology already emphasise (see, for instance, Crawford and Calo 2016; Crawford 2017; Friedman et al. 2017; Sloane and Moss 2019), needs to draw on a wider body of stakeholders, which includes those who are experts on the values and goals of a given society, such as researchers in the social sciences and humanities, and even members of society themselves. This would obviously have to be within reason, as not just any lay person will be knowledgeable, nor should they be held accountable for something over which they have no knowledge or control. However, even with bringing in the expertise of a range of people, technical researchers may still shoulder a higher degree of burden because of the fact that these researchers must acknowledge that other players need to be involved and consulted, not as a matter of courtesy or annoyance, but as central to advancing the actual research. Alternatively, organisations driving research should be required to engage a diversity of relevant experts to ensure that the epistemological sense of explicability is met, and be held accountable if they fail to do so. It is those involved in or driving the actual research who are in a position to ensure that a range of interests and values are acknowledged and, ideally, addressed in both local and international research, or that international research is not uncritically implemented across a range of differing contexts.

Ensuring that relevant experts from a range of backgrounds are engaged speaks in favour of promoting interdisciplinary research and societal engagement as a matter of necessity, not simply for ethical considerations but for advancing the research itself. Luckily, the value of interdisciplinary and multi-stakeholder engagement is already recognised in the various centres and initiatives being set up in Africa, such as the Centre for Artificial Intelligence Research (CAIR) and the South African Affiliate Centre of the C4IR. Applying a principle of explicability in an African context that recognises the necessary involvement of a range of actors, a kind of division of labour for addressing the epistemological sense of explicability, could thus generate diffuse patterns of accountability when addressing the ethical sense of explicability. What exactly this would entail in terms of recommendations, and whether accommodating a diffuse notion of accountability is feasible on the ground, however, is beyond the scope of this paper. Nevertheless, adopting and applying a principle of explicability in an African research context should aim to address these complexities.

In closing, then, we have argued that existing principles and frameworks for the development of Good AI should not be adopted uncritically into an African research context. We thus took initial steps for critically assessing one such framework, that of the AI4People report, by addressing whether the AI-specific principle of explicability should be applied in an African context. We argued that, when designing a decision-making system making use of some form of machine learning, an approach that requires adhering to a principle of explicability in both an epistemological sense (of 'how does it work?') and an ethical sense (of 'who is responsible for how it works?') not only contributes to the responsible and thoughtful development of AI that is sensitive to African interests and needs, but can also advance tackling some of the computational challenges in machine learning research. The principle thus should be adopted in an African context. Adopting the principle, however, requires that African researchers and societies, as well as organisations driving research, ensure that values are aligned, and doing so requires the involvement of a range of knowledgeable stakeholders.

Acknowledgements We would like to thank participants at the Third CAIR Symposium on AI Research and Society, held at the University of Johannesburg in March 2019, for feedback and discussion on an earlier version of the paper. We would also like to thank the two reviewers and editors for this journal for their comments.

Compliance with ethical standards

Conflict of interests Benjamin Rosman is one of the founders and organisers of the Deep Learning Indaba that we mention as an example of the growth of machine learning across Africa.

References

Abbott, J. & Martinus, L. (2019). Benchmarking neural machine translation for southern African languages. *Proceedings of the 2019 Workshop on Widening NLP*.

Andoh, C. (2011). Bioethics and the challenges to its growth in Africa. *Open Journal of Philosophy, 1*(2), 67–75.

Asilomar AI Principles. (2017). Retrieved May 29, 2019 from https://futureoflife.org/ai-principles.

Awad, E., Dsouza, S., Kim, R., Schulz, J., Henrich, J., Shariff, A., et al. (2018). The moral machine experiment. *Nature, 563*(7729), 59.

Barugahare, J. (2018). African bioethics: Methodological doubts and insights. *BMC Medical Ethics*. https://doi.org/10.1186/s12910-018-0338-6.

Beauchamp, T. L., & Childress, J. F. (2012). *Principles of biomedical ethics* (7th ed.). Oxford: Oxford University Press.

Behrens, K. (2013). Towards an indigenous African bioethics. *The South African Journal of Bioethics and Law, 6*(1), 32–35.

Biko, S. (2002). *I write what I like: Selected writings*. Chicago: University of Chicago Press.

Bostrom, N. (2003). Ethical issues in advanced artificial intelligence. In I. Smit, et al. (Eds.), *Cognitive, emotive and ethical aspects of decision making in humans and in artificial intelligence* (2nd ed., pp. 12–17). Tecumseh: International Institute of Advanced Studies in Systems Research and Cybernetics.

Brown, B. J., Przybylski, A. A., Manescu, P., Caccioli, F., Oyinloye, G., Elmi, M., Shaw, M. J., et al. (2019). Data-driven malaria prevalence prediction in large densely-populated urban holoendemic sub-Saharan West Africa: Harnessing machine learning approaches and 22-years of prospectively collected data. arXiv preprint arXiv:1906.07502.

Buolamwini, J. & Gebru, T. (2018). Gender shades: Intersectional accuracy disparities in commercial gender classification. *Proceedings of the 1st Conference on Fairness, Accountability and Transparency, PMLR*, 81, pp. 77–91.

Chukwuneke, F. N., Umeora, O. U. J., Maduabuchi, J. U., & Egbunike, N. (2014). Global bioethics and culture in a pluralistic world: How does culture influence bioethics in Africa? *Annals of Medical and Health Sciences Research, 4*(5), 672–675.

Crawford, K. (2017). The trouble with bias. *NIPS 2017 Keynote Address*. Retrieved August 19, 2019 from https://www.youtube.com/watch?v=fMym_BKWQzk.

Crawford, K., & Calo, R. (2016). There is a blind spot in AI research. *Nature, 538*, 311–313.

Deep Learning Indaba. (2019). *Together we build African AI: Outcomes of the 2nd annual Deep Learning Indaba*. Retrieved March 1, 2020 from https://www.deeplearningindaba.com/uploads/1/0/2/6/102657286/annualindaba2018report-v1.pdf.

Department of Transport. (2012). *SA learner driver manual: Rules of the road*. Pretoria: SA Department of Transport.

Doshi-Velez, F. & Kim, B. (2017). Towards a rigorous science of interpretable machine learning. Retrieved May 29, 2019 from https://arxiv.org/abs/1702.08608.

European Group on Ethics in Science and New Technologies. (2018). *Statement on artificial intelligence, robotics and 'autonomous systems'*. European Commission. Retrieved May 29, 2019 from https://ec.europa.eu/research/ege/pdf/ege_ai_statement_2018.pdf.

Floridi, L., Cowls, J., Beltrametti, M., Chatila, R., Chazerand, P., Dignum, V., et al. (2018). AI4People: An ethical framework for a Good AI Society: Opportunities, risks, principles, and recommendations. *Minds and Machines, 28*, 689–707.

Friedman, B., Hendry, D., & Borning, A. (2017). A survey of value sensitive design methods. *Foundations and Trends in Human-Computer Interaction, 11*(23), 63–125.

Gbadegesin, S. (1991). *African philosophy: Traditional Yoruba philosophy and contemporary African realities*. New York: Peter Lang.

Gyekye, K. (1987). *An essay on African philosophical thought: The Akan conceptual scheme*. Cambridge: Cambridge University Press.

Hellsten, S. (2006). Global bioethics: Utopia or reality? *Developing World Bioethics, 8*(2), 70–81.

House of Lords Artificial Intelligence Committee. (2018). *AI in the UK: Ready, willing and able?* UK Parliament. Retrieved May 29, 2019 from https://publications.parliament.uk/pa/ld201719/ldselect/ldai/100/100.pdf.

IEEE Global Initiative on Ethics of Autonomous and Intelligent Systems. (2019). *Ethically aligned design: A vision for prioritising human well-being with autonomous and intelligent systems* (first edition). Retrieved May 29, 2019 from https://ethicsinaction.ieee.org.

Kroll, J. A. (2018). The fallacy of inscrutability. *Philosophical Transactions A, 276*, 20180084.

Landhuis, E. (2017). Tanzania gears up to become a nation of medical drones, *npr.org*. Retrieved February 28, 2020 from https://www.npr.org/sections/goatsandsoda/2017/08/24/545589328/tanzania-gears-up-to-become-a-nation-of-medical-drones.

London, A. (2019). Artificial Intelligence and black-box medical decisions: Accuracy versus explainability. *Hastings Center Report, 49*(1), 15–21.

Mbembe, A. (2001). *On the postcolony*. Berkeley: University of California Press.

Mbiti, J. (1990). *African religions and philosophy* (2nd ed.). Portsmouth: Heinemann.

Microsoft. (2019). *Artificial Intelligence for Africa: An opportunity for growth, development and democratisation*. Retrieved May 29, 2019 from https://info.microsoft.com/ME-DIGTRNS-WBNR-FY19-11Nov-02-AIinAfrica-MGC0003244_01Registration-ForminBody.html.

Montreal Declaration for a Responsible Development of Artificial Intelligence. (2017). Retrieved May 29, 2019 from https://www.montrealdeclaration-responsibleai.com/the-declaration.

Moodley, K. (2007). Microbicide research in developing countries: Have we given the ethical concerns due consideration? *BMC Medical Ethics*. https://doi.org/10.1186/1472-6939-8-10.

Murove, M. F. (2005). African bioethics: An explanatory discourse. *Journal for the Study of Religion, 18*(1), 16–36.

Partnership on AI. (2018). *Tenets*. Retrieved May 29, 2019 from https://www.partnershiponai.org/tenets.

Rakotsoane, F., & Van Niekerk, A. (2017). Human life invaluableness: An emerging African bioethical principle. *South African Journal of Philosophy, 36*(2), 252–262.

Ramose, M. (2005). *African philosophy through Ubuntu* (Revised ed.). Harare: Mond Books Publishers.

Selbst, A. D., & Powles, J. (2017). Meaningful information and the right to explanation. *International Data Privacy Law, 7*(4), 233–242.

Senghor, L. (1988). *Ce Que Je Crois*. Paris: Grasset.

Sloane, M., & Moss, E. (2019). AI's social sciences deficit. *Nature Machine Intelligence, 1*, 330–331.

Steyn, L. (2018). Watch your job, the bots are coming. *Mail & Guardian*, 8 December. Retrieved May 29, 2019 from https://mg.co.za/article/2017-12-08-00-watch-your-job-the-bots-are-coming.

United Republic of Tanzania Ministry of Health and Social Welfare. (2015). *Health Sector Strategic Plan July 2015-June 2020*. Retrieved February 28, 2020 from https://www.moh.go.tz/en/strategic-plans.

Wa'Thiongo, N. (1986). *Decolonising the mind*. Chowchilla, CA: James Curry.

Wachter, S., Mittelstadt, B., & Floridi, L. (2017). Transparent, explainable, and accountable AI for robotics. *Science Robotics*. https://doi.org/10.1126/scirobotics.aan6080.

Winfield, A. F. T., & Jirotka, M. (2018). Ethical governance is essential to building trust in robotics and artificial intelligence systems. *Philosophical Transactions A, 376*, 20180085.

World Health Organisation (WHO). (2020). Guidelines on reproductive health research and partners' agreement. *Sexual and Reproductive Health*. Retrieved February 21, 2020 from https://www.who.int/reproductivehealth/topics/ethics/partners_guide_serg/en/.

Publisher's Note Springer Nature remains neutral with regard to jurisdictional claims in published maps and institutional affiliations.

Ethics and Information Technology (2021) 23:119–126
https://doi.org/10.1007/s10676-020-09529-z

ORIGINAL PAPER

An ontic–ontological theory for ethics of designing social robots: a case of Black African women and humanoids

M. John Lamola[1] ⓘ

Published online: 29 February 2020
© Springer Nature B.V. 2020

Abstract

Given the affective psychological and cognitive dynamics prevalent during human–robot-interlocution, the vulnerability to cultural-political influences of the design aesthetics of a social humanoid robot has far-reaching ramifications. Building upon this hypothesis, I explicate the relationship between the structures of the constitution social ontology and computational semiotics, and ventures a theoretical framework which I proposes as a thesis that impels a moral responsibility on engineers of social humanoids. In distilling this thesis, the implications of the intersection between the socio-aesthetics of racialised and genderised humanoids and the phenomenology of human–robot-interaction are illuminated by the figuration of the experience of a typical black rural African woman as the user, that is, an interlocutor with an industry-standard socially-situated humanlike robot. The representation of the gravity of the psycho-existential and socio-political ramifications of such woman's life with humanoids is abstracted and posited as grounds that illustrate the imperative for roboticists to take socio-ethical considerations seriously in their designs of humanoids.

Keywords Computational semiotics · Humanoids · Robot gender · Robotic ethics · Robot race · Postphenomenology

Introduction

The objective of this paper is to contribute a theoretical framework which demonstrates why it is imperative that ethics, and in particular social ethics, should be taken into consideration in the design of humanoids. Working from the general context of the philosophy of technology, Ihde (1990, pp. 141–143) and Verbeek (2005, pp. 125–146) have alerted of the technological intentionality that is at play during human–machine-interaction, and how this affects the user's sense of being in the world. I take this insight further into the specific case of socially situated humanlike robots in which I polemically perceive the latter in their culturally-influenced socio-aesthetic state, that is as objects of knowledge (the ontical) which peculiarly bear the potential to frame our psychic-existential state (the ontological). I explore and demonstrate how this human vulnerability to technological intentionality with its commensurate ontological phenomenology obligates ethical responsibility in the building of robots.

Venturing onto the socio-ontological, Verbeek unwittingly, in my assessment, underscored that "when technologies are used, they co-shape human-world relationships: they make possible practices and experiences, and in so doing, they play an active role in the way humans can be present in their world and vice versa" (2005, p. 140). Our disquisition is framed around a normative injunction which could be drawn from Verbeek's observation when strictly applied to the problematique of the consequences of engineering design-decisions on the dynamics of human–robot-interaction. Such axiology is exemplified by the following tenet from the mission statement of *The Hague's Foundation for Responsible Robotics*, which cautions that, "in robots, we not only project who we are but we come to affect who we will become" (Sharkey et al. 2017, p. 42).

My critical originary point is that a robot is a product of human ingenuity and labour; therefore, a humanoid social robot is quintessentially a cultural artefact. As a human-like robot that is adorned with anthropomorphic features, it compositely reflects the preferences, assumptions and prejudices of the software programmers, the robotic engineers and the

✉ M. John Lamola
 jlamola@mweb.co.za

1   Department of Philosophy, University of Pretoria, Pretoria,
    South Africa

Chapter 3 was originally published as John Lamola, M. Ethics and Information Technology (2021) 23: 119–126. https://doi.org/10.1007/s10676-020-09529-z.

financial interests that go into a robot-building project. The ultimate technological output embodies the cultural[1] totality from which the design of the humanoid is derived, or the design-decision idiosyncrasies of its engineers.[2] A humanoid, as such, is a technological product, which qua technology, it is neither culturally generic nor politically innocuous.

Grounded upon this "first truth", I explore insights from the fields of phenomenology and semiotics which indicate the intricate manner in which humans are affected by, or react towards these humanlike artificial agents. This provokes my thesis that ethical considerations should govern the engineering of humanoids as the latter tend to assume an ontology that traverses between the boundaries of technology and the epistemological protocols of human cognition, eliciting in the process psychical reactions that potentially have socio-psychological ramification on its users. In amplifying my point, I adopt and deploy the case of black women in Africa as the human figure[3] which is in a socio-technological intercourse with humanoid robots. The socio-existential condition of the black African female subject on the African continent,[4] typified here as being rural, is arguably that of an existence at the bottom stratum of the global hierarchy of access to the benefits and social power dynamics of the so-named Fourth Industrial Revolution. Around this *Figure*, the issues of gender, social class, race, aesthetics and disparities in global digital equity as pertaining to robotics are conglomerated and symbolised.

In this context, while noting discursive contours on feminist theories explored in Africanist writings such as those by Osha (2008) and contributors to Basu (2018), the socialist-feminist sensibilities raised by Donna Haraway in "A Manifesto for Cyborgs" on the challenges occasioned by technoscience on female subjects are most directly pertinent (in Cahoone 2003, pp. 464–478). Although within the constraints of this article I posit the black African woman as a *figuration*, a case of thought, in service of my advocacy for politico-aesthetically conscious robotic designs, I firmly

affirm, to use Haraway's idiom, that the African woman's "physicality is undeniable" and is taken throughout our reflection as "deeply historically specific" (Haraway in Ihde and Silinger 2003, p. 49). The reality of genderalisation of humanoids and its social consequences is so serious that in her *Robo Sapiens Japanicus: Robots, Gender, Family, and the Japanese Nation* Robertson (2017) finds it necessary to instructively emphasise that "some humanoids are so lifelike that they actually pass as human beings. These gendered robots are called *androids* (male) and *gyroids* (female)" (Robertson 2017, p. 6). Clinically, we should not use the word "humanoid" without specifying its sexual orientation.

In their research paper "Persuasive Robotics: The influence of robot gender on human behaviour", Siegel et al. (2009) demonstrated how sexual-ascription, colour, bodily shape and hair-type of a robot are factors critical to the user's behavioural mode of interaction with their robot. This observation on the psychical effect of the socio-aesthetics of humanoids was corroborated by the *Robots and Racism* report to the American Institute of Electrical and Electronics Engineers (IEEE) 2018 annual conference by Christoph Bartneck and his multinational research team (Bartneck et al. 2018 herein after, "the Barneck Report"). Noting that "because race corresponds with complicated patterns of social relationships, economic injustice, and political power, the perception of race in the design space of robots has potential implications for HRI" (p. 196), Barneck et al. set out to ask: "do people automatically identify robots as being racialized, such that we might say that some robots are 'White' while others are 'Asian' or 'Black', and are there socioethical concerns therein?" (ibid.). In pursuit of this question, the research team conducted an extended replication of the classic social psychological shooter bias paradigm gauging human reaction to robot stimuli. They found that: "Reaction-time based measures revealed that participants demonstrated 'shooter-bias' toward both Black people and robots racialized as Black. Participants were also willing to attribute a race to a robot "on the basis of their racialization and demonstrated a high degree of inter-subject agreement when it came to these attributions" (Bartneck et al. 2018, p. 197).

Having underscored their empirical report's conclusion on how people impute racial and genderised identities onto robots with a refrain that "there is no need for all robots to be white" 2018, p. 197), the Barneck team published a sequel to their report with a journal article with an evocative Design Ethics title: "Robots Can Be More Than Black And White" (Addison et al. 2019).

Against this background that draws attention to the proclivity and potential effects of discriminatory representations in the production of robots, the goal of this paper is to support what Harris et al. in *Engineering Ethics: Concepts and Cases* idealise as "becoming a socially conscious engineer"

---

[1] The use of and meaning of "culture" implied here transcends the conception of culture as an ethnic practice. It extends to the composite stage of intellectual-epistemological practices and norms of a given society and even a Civilisation.

[2] *Ted quim*, it is remarkable how the makers of the famed Sophia robot avoided to adorn "her" with hair in order to obviate ethnic connotations, and only did so in a much publicised occasion of the debut of this humanoid on China's national CTV programme in Beijing in January 2018. See https://chinaplus.cri.cn/photo/china/18/20180 115/78288_3.html.

[3] Our usage of 'figure' is derived from its formulation in Deleuze as *Figure* or figuration (2003, pp. 1–11), and as utilised in Haraway (2003, pp. 48, 49).

[4] The specificity of the descriptor "on the African continent", besides being deployed to maximally illustrate the element of global digital disparities, is in part inspired by Atanga (2013).

(Harris et al. 2009, p. 91). I propose to contribute to robotic technology design protocols what Jacobs and Huldtgren (2018) call "value sensitive design" by providing a philosophical paradigm that maps out the theoretical terrain upon which a veritable ethical obligation can be grounded. My mission is to present collateral theoretical content for the conscientisation of engineers and related participants in the science and business of the design and building of humanoids. The discussion is framed for dialogue with innovative professionals who are responsible for the design-decisions that perambulate the race and gender of a robot: the features that frame the robot's sociality, and ultimately, the counter-ontological effects on its human interlocutor.

I begin with a philosophically nuanced rendition of the epistemological status of humanoids which distinguishes the latter from the general hubris of robotics technologies. Flowing from this phenomenal portrait of socially-situated robots, I introduce and elucidate the role of semiotics as a processes of meaning communication which I apply to the dynamics of human–robot-interaction. The semiotic, phenomenological and psychical suggestive influences of robots which we highlight, are then, towards my conclusion, interpreted as an ontic–ontological proposition, that is, a theory that explicates how the process of whatever we come to know relates to social existence. This proposition, which is a thesis that I develop systematically throughout the paper, is progressively deployed as a corroboration for social ethics to be taken much more seriously in the engineering of humanlike robots.

Status of humanoids as epistemic objects

From the perspective of epistemology, a robot, an industrial robot in particular, whilst a technological artefact of curious wonder, is an object like any other object impressed onto our minds via our sense of sight. In Kantian epistemology it is *noumena*, an epistemic (knowable) object. On the other hand, a robot which is expressly designed and presented as a human companion or caregiver, although recognised as an artificial agent, imposes some special salience to our senses. It does not impress itself onto our cognitive space like any other epistemic (knowable) object, say, a chair or even a painting of a human face. As a technological artefact with human-like behavioural traits that are geared at performing typically human roles, endowed with features of the functionality of the human bio-neurological system, it resembles us; it is a *humanoid*. At the same time, it is a uniquely humantological technology: it provokes immediate cognitively implicit as well as philosophically explicit questions about what is human.

As a socially-situated artefact, a *social* robot, it shares our space in a socially affective manner. It simulates human existence as existence-in-community-with others; it traverses into Aristotle's *Homo politicus*. As such, we *experience* it

as both a representation and part of our Being. In Martin Heidegger's terms, we are *Mitsein* (being-with) (Heidegger 1962/1927, p. 157) with social robots. Strasser's assertion is apt: "Where previous [scientific] revolutions have dramatically changed our environments, this one has the potential to substantially change our understanding of sociality" (2017, p. 106).

The foregoing philosophic statements, which I will explicate as we proceed, have two crucial implications. The first is that the *existence* of social humanoid robots, which comes into being through the manufacturing or private acquisition of a robot, including a casual encounter, with such human-like robot, adumbrates the instantiation of a socio-technical world (*umwelt*), the "living with robots" (See Doumouchel and Damiano 2016). Secondly, which is of our immediate interest at this stage, is that, as a robot, appearing to us in its aesthetic and behavioural semblances that mimic human life, it readily locks us into a phenomenology of intricate intersubjectivity. We momentarily experience it as (or believe it to be?[5]) a human. This moment of human-humanoid encounter is devoid of cognitive dissonance. A realisation of this absorption or self-wrapping of the human mind around a social robots supports remonstrations of postphenomenologists such as Kelly's (2015) against Edmund Husserl, pioneer of transcendental phenomenology.[6] Kelly elucidates that "when technology is introduced, both the human experiencer and the thing experienced are transformed" (2015, p. 508). He then asserts that "at this interrelational level, technologies may be more than just another object in the world of which the human experiencer is conscious [...] (ibid.).

More than any other technological artefact or tool, which is the generic "technology" that Kelly (2015, p. 508) is dealing with in his argument that classic phenomenology should appreciate the possibility of intersubjectivity with non-human objects, a robot simulating human existence provokes an exaggerated attention, curiosity and interest. A recent research experiment by Cao et al. found that children with Autism Spectrum Disorder (ASD) engaged in more eye contact and were fixating for longer periods on a humanoid robot face that on a human face (Cao et al. 2019). This point signifies the difference within the domain of HRI between human–robot-*interface*, in which "the robot" may be an industrial tool or intelligent machine, and human–robot-*interaction* or interlocution, in which "the robot" is specifically some human-like artificial agent.

---

[5] Marti et al. (2006) and Pollini (2009, p. 169) introduce the concept of "a suspension of belief" as a dynamic at play when humans cognitively encounter humanoids, an act similarly observed in a toddler playing with dolls.

[6] For a lucid introduction to the thought of Husserl as the foundation upon which Martin Heidegger and Jean-Paul Sartre developed their phenomenological methods, see Moran (2000, pp. 1–20, 60–90).

This curiosity that is induced by humanoid robots is in part similar to the human proclivity of finding other humans interesting. They "catch our attention", even on photographs. They exert this particular effect because they resonate our *selves*. Similarly, a humanoid robot, be it an assistive of aggressive-like Robocop, reminds us of our selves, at least, our humanity; we re-*cognise* something about us in them, and subconsciously expect them to act like or with us. As an illustration of this "subconscious reminding" Romesin and Bunnel (1998, p. 34) invite us to think of a suburban mother walking her little girl in a park whereupon they stumble onto two dogs copulating. The mother smirks and implores the daughter, "Don't look at them!" Why? Because the dogs remind her of what we adult humans do only in private spaces. This bears an analogous similarity to the phenomenological process of human–robot-interaction. They may not, according to our prior knowledge, be human, but they affectively provoke humanistic expectations from us.

A human-like robot, therefore, proves that it is not a banal epistemic "object" of a phenomenologically active mind, that is, a phenomenon in the sense a phenomenologist like Berghofer (2019) would explicate Husserlian noetics. It has epistemological agency, which is peculiarly akin to that discernible in human–human interaction. It is an artefact which is autonomously imbued with meaning-emitting value. As a humanoid, it is an artefact expressly engineered with a semiotic intention. It is directed at signifying something other than itself, as its simulacra: a living human being. It is an embodied, embrained (software encoded) and encultured image fashioned for self-representation to the human mind as something-like-human, and for consumption (receptivity) by the mind as such. Without this lexical cognitive-epistemological value chain, there would be no social robotics industry. The drive for innovation in social robotics is premised on an aspiration to progressively design and produce robots that optimally look and feel like human to the human mind, as this is the prime condition for their successful marketing and social deployment. As such, a humanoid is an image/representation with an absolute semiotic ontology; it is meant for meaning, and the emission of this meaning as impressed in, and apprehended by the human brain constitutes the humanoid's epistemological agency.

In the context of the study of signs, a sign should be subordinated to what it is representing, wherein the signified or referent is primary to the signifier. However, we notice that as a semiotic apparatus, the robot masterfully endowed with sociality is a *sign* of a peculiar order. It is *meant* (by its manufacturers) to look, and it does look like what it is meant to represent or signify (otherwise it is a failed project). Within a typology of semiotic representations, it is an icon. Alas! in virtue of it being a humanoid, a robot supposed to be a *referee* of the human *referent*, it is not dislocated from its referent; the direction to the referent is embedded within the humanoid robot; it *is* the referent, as its very essence and ontic value is to simulate the human person with lexical perfection. Its success in attaining the ontology or "commercial" status of *the social robot*, is its quality of being a near-perfect representation and resemblance of a human person as a composite display of an active neural system and physiological features.

As a corollary stage of our reflection that should further corroborate the qualitative claims I make in the foregoing, we have to proceed into a closer interrogation of the affective nature and sociological status of these meanings that are emitted by our encounter with these socially-situated and culturally-designed robots.

Robotic persuasiveness and its ramifications

Rightly named, humanlike social robots are *persuasive humanoids* as they provoke both ontic-illusory[7] and intuitive mental acts from their human users or encounterors. As variously researched and reported, for example in Coeckelbergh's "Why Care about Robots" (2018), an image represented by a humanoid exerts a suggestive influence, both conscious and unconscious, on the user's mental state. Amongst his variety of examples, Coeckelbergh relates the case of HitchBOT, a 2014 Canadian robot which as programmed, successfully hitchhiked in a number of countries, including the United States of America "with the help of friendly strangers",[8] according to the press release of the HitchBOT project team. "Its" journey was tracked on social media platforms by an ever-widening community of fans. HitchBOT's journey ended abruptly on 1st August when he was found vandalized in Philadelphia: "his" head and arms ripped off. Even though morphologically, by robotic design standards, this was not a perfect humanoid, its damage provoked an outpouring of empathetic emotions on social media, with one fan moaning: "America should sit in the corner and think about what it's done to poor HitchBOT" (Coeckelbergh 2018, p. 142).

When a humanlike robot appears to our senses, we experience it, and make sense of it; we get an *impression* of it. This process of making sense, and the eventual meaning-given, reflects affectively on the encounterer/encountered; the robot *expresses* "itself" (reveals itself?) to our cognitive faculties. This expression-impression dynamic, in Husserlian terms, is the meaning-making dynamic, the attainment of the object-as-intended, the *noema* (Husserl 2008/1906, pp. 17–20). But the Husserlian rendition is not adequate as

---

[7] By "ontic-illusory" I seek to denote, the immediate suggestion at an initial point of cognitive encounter that an object could be something which is not what it is, but which vision my mind overrides.

[8] https://mir1.hitchbot.me/.

it plays down the full force of the expressing object. My consideration, as outlined above on the autonomous semiotic agency of a humanoid, is that an object of consciousness, here specifically the socialised robot, is itself actively imposing its semiotic ontology into our cognitive space. It is not a mere anthropomorphic illusion. I will endeavour to explain this further in terms of Hegel's dialectic phenomenology. The meaning of the robot as my ultimate apprehension of what it is, is its autonomous self-reconciliation to my mind as the subject, its transcendence (*aufhebung*) of its momentary alienated (unclear) state as an object.

In a humanoid, I *re-*cognise something like me. In (Hegelian) phenomenological terms, I understand/comprehend (*begriff* not *verstehen*) it as something like, or posing to be me. In semiotic terms, the robot is a signifier in which the signified is the mirror image of myself as the represented (reflected/*representation* of) human being. This human-being signification is for this very reason human-affective, it affects me psychologically as would a real human being. Hence, for an easy example, we have the case of an encounter between a feminised sex-robot (gyroid) and a sexually active heterosexual male. He gets an erection. He is *impressed* by the *expression* conveyed in and by this sexualised artefact.[9]

Both the phenomenology of how social robots remind us of ourselves, and the nature of the semiotic character of these artificial-human-like artefacts bear far-reaching consequences for our human ontology and ethics. Adding on what we found about the humanoid robot in the foregoing, Pierce precociously defined a *sign* in 1908 as "anything which is so determined by something else, called its object, and so determines an effect on a person, which effect I call its interpretant, that the latter is thereby immediately determined by the former" (Pierce 1960/1908, p 48). Ironically, the latent possible design-ethics ramifications of the human–robot semiosis we are preoccupied with here in our twenty-first century robotics study are suggested in this canonical script. Here ontological-phenomenology, semiosis and ethics converge. This convergence is best demonstrated in the prevalent consensus, aptly articulated by Coeckelbergh, that "mistreating a robot is not wrong because of the robot, but because doing so repeatedly and habitually shapes one's moral character in the wrong kind of way" (2018, p. 145). The affectivity and the mode of the regard of a humanoid ricochets into shaping who we become.

What are the probabilities of the ethnic features and sex of a robot replicating the social status that mirrors the role typically imposed to that particular racial group or gender in a racialised and patriarchal society? Are the digital voice assistants in our computer devices such as Amazon's Alexa and Microsoft's Cortan female because women are typically "assistants", polite and efficient secretaries?[10]

Recalling our originary point on how a robot as a product of human ingenuity and labour is quintessentially a cultural artefact that reflects the design preferences and proclivities of its creators, and the results of the Siegel (2009) experiment on the psychological effects of the gender of robots, together with the Bartneck et al. (2018) lament of the standard industry practice of designing humanoids as white, we are directed to a much deeper question when we take into account how this persuasive effect of humanoids modulate human behaviour not only towards the robots, but also human self-perception or self-image vis a vis the robot.

This "deeper" question presents itself as some form of an ontological-existential crisis. Verbalised by Jackie Snow, the humanoids Siegel and Bartneck refer to as technocultural artefacts, "look like the people building them, but not necessarily using them".[11] In politically charged culturo-aesthetic terms, these robots are racialised white, and according to their symbolic rationality, are male.[12] They are *persuasive white androids*; but they are used, that is, interact, with people who may not be male and white. How would this affect those who interact with these robots? In our case, we hypothetically posit that they are used by a black woman in some rural locale on the continent of Africa. What would be immediate social power dynamics between her and such a white android?

Linda Martin Alcoff wrote that "in much feminist literature the normative, dominant subject position is described in detail as a white, heterosexual, middle-class, able-bodied male" (1998, p. 8). On the other hand, in her "Manifesto of Cyborgs" Haraway challenges the maleness of post-Second World War information science, accusing it of "phallogocentrism" (Haraway 2003, p. 475). Is it possible that a humanoid robot may have an oppressive/discriminative effect on me in virtue of a genderised mental attitude it is programmed with (male phallogocentrism), or the racial physiologicalities in which it is cast in the context of a racialised society?

---

[10] Instructively, in an apparent response to ethical sensibilities similar to what we alert in this paper Apple (Alphabet) upgraded its Siri to perform as either a male of female voice.

[11] Jackie Snow in https://www.fastcompany.com/90212508/even-black-robots-are-impacted-by-racism [Accessed 2019-06-03].

[12] On the maleness of Reason, see Borno (1986). I further aver that this Western Cartesian mode of rationality is replicated in computer languages and artificial intelligence.

The developing import of our present dissertation is a suggestion that the way a robot looks and behaves may affect the self-image of the user or entrench certain patterns of human social relationality; a dimension of this reflection could be on how this proceeds to affect the very existential self-knowledge of the encountering human being as a socially located agent, that is one's social ontology.Taking our case of the black female subject in Africa, we could then claim, a posteriori, that social robots have the potential of perpetuating a black African feminine existentiality of self-marginalisation, socio-economic abjection and techno-exclusion.[13] When abstracted away from our case of the African woman these observations, of their own theoretic merit, buttress the importance of value-sensitive engineering designs that cohere with one of the key stipulations of the *Asilomar Principles on Research on Artificial Intelligence* that "AI systems should be designed and operated so as to be compatible with ideals of human dignity, rights, freedoms, and cultural diversity".[14]

But a rebuttal could be posed against my hypothesis that the physiological morphology and the aesthetics that socially stereotype the look of a humanoid may have an inverse effect of framing the human interlocutor's self-image, that is, her existentiality. This would be the question: what about human agency, and the intentionality that could be deployed to counteract the deleterious suggestive influences of a human-like robot? This is a matter I now turn to as we approach our concluding section.

Absolutist computational semiotics and its ramifications

Indicating the semiotic dimension of computing languages, the eminent semiotician Umberto Eco, for one, reminds that in their semiosis images and words have an inherent problem of susceptibility to a variety of interpretations and hermeneutic appropriations that are contingent, amongst others, on the cultural and ideological positionalities of their interpreters (Eco 1997, pp. 174, 308). On this basis, it may therefore be assumed that one may randomly either be negatively or positively *impressed* by what is being expressed by a social robot, depending on the function of their apprehensive involvement, that is (eiditic) intentionality.

In contrast to classic semiotics, robotic or computational semiotics vitiates the ambiguities that Eco alludes to; it departs from the point of the lexicality of the image

represented by the robot (see Gudwin and Queiroz 2005). As we noted, the successful production of a humanoid robot is determined by the degree at which it resembles and mimics human existence and roles. The robotic image is in this instance self-definitional. It is as singular as a road sign with the word "Stop" inscribed on it. One cannot separate the indicated message from the sign. Such a sign can be contrast to a directional sign that gives information that points away from itself. The philosophy of the science of social robotics is that the intended robot is not an image denoting something else, but seeks to equate as much as possible what is represented.

Uniquely, and unlike in hermeneutic semiotics, computational semiotics is premised on an assumption or imposition of universal comprehensibility; that the viewer, user or interactor will immediately be satisfied that the image or robotic artefact is the mimicry of a human person. The meaning of what is represented is expressly disambiguated; it is a humanoid, and not some dog or ape (animaloid). There is no polysemiosis with the socially-situated humanlike robot; it strives to convey an image of a real human person in a predetermined role. A semiotic convention is at play here. This is peremptorily declared by Eco: "Computer languages... are universal systems; they are comprehensible to speakers of different natural languages and are perfect in the sense that they permit neither error nor ambiguity" (Eco 1997, p. 311).

As explicated in Clark and Chalmers (1998) in their paradigmatic article "The Extended Mind", the modern human mind is already immanently interwoven with the functionalism of computers. This fusion of the human and technology world is of late demonstrated by the burgeoning field of Internet of Things (IoT), of which robotics is its most advanced expression (Royakkers et al. 2018, p. 127). This ubiquitous "language" of machines, modelled to replicate the structure of the human mind, Clark and Chalmers have proposed, is in fact a shared and integrated human–computer "mentalese" (1998, p. 7). Similar to Eco's view of computer language, this mentalese, as presented as the coupling of the external cognitive impulses from an artefact with the human internal cognitive process, is according to Richard Melany universal, and self-imposing in its syntactics and semantics (Menary 2010, p. 207).

Linked to what we noted earlier on the kind of a sign that a humanoid robot is, that it is an icon in which the *referent* is subsumed into the *referee*, and thus rendering a humanoid an absolute semiological ontology, we now note the absolutist peculiarities of computational semiotics outlined above as the pervasiveness of human–machine mentalese. I can therefore claim that I am justified by theory in assuming an occurrence of a possibility of a univocal connotation or impression of what may be a negative persuasion/influence toward interactors with a humanlike-robotic output, as

---

[13] I expounded on this subsequent conclusion in a paper presented at the Research Colloquium of the University of Fort Hare, South Africa on 7th May 2019 "Black Women and Robots: A Propaedeutic Reflection on Artificial Intelligence and African Existentiality".

[14] https://futureoflife.org/ai-principles/?submitted=1#confirmation [Accessed 2019-09-28].

dependent on the socio-aesthetic anthropomorphic features a social robot is designed with.

Moreover, it is particularly noteworthy in the context of our discussion that in asserting that computer languages are "universal systems", Umberto Eco concedes that "their rules are drawn from the western logical tradition" (1997, p. 311). These rules are what Jacque Derrida in *White Mythology* derided as "logocentrism" (Derrida 1974, p. 7), which is the root term of what Haraway excavated as "phallogocentrism". Mentalese is "white, heterosexual, middle-class, able-bodied male" to borrow Alcoff''s encryption (1998, p. 8).

When the power of computational semiology is paired with insights from post-phenomenology, it evinces dramatic implications for a user such as our black African female subject. The postphenomenology movement in the philosophy of technology holds that in the meaning-making process that mediates human-technology interaction, the object, the technological artefact, must be accorded a rehabilitative privileged position over the perceiving subject (Ihde 2003, pp. 131–144; Roosenberger and Verbeek 2015; Tripathi 2017, pp. 137–148). The case of the black African woman as socio-economically positioned in the global power matrix at the bottom of the pile in a socio-technological episteme that privileges the semiotic power and the prior status of the robot object to that of her as a phenomenological subject, raises a serious ethical obligation on the designers of humanoids. What is the socio-existential status of a woman in rural Somalia, *vis a vis* that of the humanoid Sophia who was recently granted citizenship of the oil-rich Saudi Arabia?[15] Besides this *woman*-to-woman comparison, could phallogocentrist robots, programmed with the neural architecture of the western logical tradition and aesthetic features that affirm white male ontological normativity turn out to be absolute oppressors (with techno-semiotic permanence) of African women?

Conclusion

In my endeavour at constructing a theoretical system that is derived from philosophical traditions that deal with meaning-making in the context of technology and the formation of social ontology, I have developed a novel appreciation of the aesthetic and phenomenal ontology of humanlike socially-situated robots. This in turn has served to account for the affective potentialities of the latter. I have foregrounded the socio-political issues of race and gender as pertinent factors in design decisions, impelling roboticists to be more conscious of their cultural and ethnic positionalities,

and perhaps even their political commitments. In order to focus on delivering the structure of the theoretical account I introduce here, I have not been able to delve into the details of the psycho-existential, socio-ontological and political ramifications of how social robots as technological outputs as finally delivered could affect the existentiality or self-consciousness of individuals who are placed in the global techno-economic power matrix in a position such as that of black African women in rural Africa. I merely highlighted this as a case of thought. The experience of humantologised technology by this *Figure* has been demonstrated as placing crucial ethical obligations on engineers of humanoids, and social robots specifically.

References

Addison, A., Yogeeswaran, K. & Bartneck, C. (2019). *Robots Can Be More Than Black And White: Examining Racial Bias Towards Robots: In Proceedings of Conference on Artificial Intelligence, Ethics, and Society.* Honolulu: Association of Advancement of Artificial Intelligence www.aies-conference.com/wp-content/papers/main/AIES-19_paper_99.pdf. Accessed 16 June 2019.

Alcoff, M. L. (1998). What should white people do? *Hypatia, 13*(3), 6–26.

Atanga, L. (2013). African feminism. In L. Atange, S. E. Ellece, L. Litosseliti, & J. Sunderland (Eds.), *Gender and language in sub-Saharan Africa* (pp. 301–314). Amsterdam: John Benjamin Publishing.

Bartneck, C., Yogeeswaran, K., Ser, Q. M., Woodward, G., Sparrow, R., Wang, S., & Eyssel, F. (2018). *Robots and racism: Proceedings of ACM/IEEE International Conference on Human-Robot-Interaction.* Chicago: IEEE. pp. 196–204. doi: 10.1145/3171221.3171260

Basu, A. (Ed.). (2018). *The challenge of local feminisms: Women's movements in global perspective [1995].* New York: Routledge.

Berghofer, P. (2019). Husserl's Noetics: Towards a phenomenological epistemology. *Journal of the British Society for Phenomenology, 50*(2), 120–138.

Bordo, S. (1986). The cartesian masculinization of thought. *Signs Journal of Women in Culture and Society, 11*(3), 439–456.

Cao, W., Song, W., Li, X., Zheng, S., Zhang, G., Wu, Y., et al. (2019). Interaction with social robots: Improving gaze toward face but not necessarily joint attention in children with autism spectrum disorder. *Frontiers of Psychology, 10*, 1503. https://doi.org/10.3389/fpsyg.2019.01503.

Clark, A., & Chalmers, D. (1998). The extended mind. *Analysis, 58*(1), 7–19.

Coeckelbergh, M. (2018). Why care about robots? Empathy, moral standing, and the language of suffering, Kairos. *Journal of Philosophy & Science, 20*, 141–158. https://doi.org/10.2478/kjps-2018-0007.

Deleuze, G. (2003). *Francis Bacon: The logic of sensation* (Daniel W. Smith Trans.). London: Continuum. (Original work published 1981)

Derrida, J. (1974). *White mythology: Metaphor in the text of philosophy [1971].* Baltimore: John Hopkins University.

Doumouchel, P. & Damiano, L. (2016). *Living with robots* (M. Bevoise Trans.) Cambridge MA: Harvard University Press.

Eco, U. (1997). *The search for a perfect language.* Oxford: Blackwell Publishing.

Gudwin, R. & Queiroz, J. (2005). *Towards an Introduction to computational semiotics: Proceedings of the International Conference on*

---

[15] https://www.biztechafrica.com/article/sap-africa-brings-humanoid-robot-mzanzi/13892

 Springer

Integration of Knowledge Intensive Multi-Agent Systems (KIMAS 18–21 April 2005, Waltham, MA, USA). Chicago: IEEE.

Haraway, D. (2003). A manifesto for cyborgs: Science, technology, and socialist feminism in the 1980s. In L. Cahoone (Ed.), *From modernism to post modernism* (2nd ed., pp. 464–480). Oxford: Blackwell.

Harraway, D. (2003). Interview with Donna Haraway with R. Markussen, F. Oesen and N. Lykke. In D. Ihde & E. Selinger (Eds.), *Chasing Techscience: Matrix for materiality* (pp. 47–57). Bloomington: Indiana University Press.

Harris, E. C., Prichard, S. M., & Rabins, J. M. (2009). *Engineering ethics: Concepts and cases* (4th ed.). Belmont, CA: Wadsworth.

Heidegger, M. (1962). *Being and time* (J. Macquarie & E. Robinson, 7 Ed, Trans.). Oxford: Blackwell. (Original work published 1927)

Husserl, E. (2008). *Introduction to logic and theory of knowledge: Lectures 1906/07* (C. O. Hill, Trans.). Dordrecht: Springer.

Ihde, D. (1990). *Technology and the lifeworld: From garden to Earth.* Bloomington: Indiana University Press.

Ihde, D. (2003). If phenomenology is an Albatross, is post-phenomenology possible? In D. Ihde & E. Selinger (Eds.), *Chasing Techscience: Matrix for Materiality* (pp. 131–144). Bloomington: Indiana University Press.

Jacobs, N., & Huldtgren, A. (2018). Why value sensitive design needs ethical commitments. *Ethics and Information Technology.* https://doi.org/10.1007/s10676-018-9467-3.

Kelly, M. R. (2015). Edmund Husserl. In J. B. Holbrook & C. Mitcham (Eds.), *Ethics, science, technology, and engineering: A global resource* (2nd ed., Vol. 4, pp. 507–509). Farmington Hills MI: Cengage Learning.

Marti, P., Bacigalupo, M., Giusti, L., Mennecozzi, C., Shibata, T. (2006). *Socially assistive robotics in the Treatment of Behavioural and Psychological Symptoms of Dementia: Proceedings of the BioRob 2006: IEEE/RAS-EMBS International Conference on Biomedical Robotics and Biomechatronics.* Pisa, Italy.

Menary, R. A. (2010). Cognitive integration and the extended mind. In R. A. Menary (Ed.), *The extended mind.* Cambridge, MA: MIT Press.

Moran, D. (2000). *Introduction to phenomenology.* London: Routledge.

Osha, S. (2008). Philosophy and figures of the African female. *Quest An African Journal of Philosophy/Revue Africaine de Philosophie, 20,* 155–204.

Peirce, C. S. (1960). Collected papers of Charles Sanders Peirce. In C. Hartshorne & P. Weiss (Eds.), *Principles of philosophy [1908].* vol 1, Cambridge, MA: Harvard University Press.

Pollini, A. (2009). A theoretical perspective on Social Agency. *A. I and Society, 24*(2), 165–171.

Robertson, J. (2017). *Robo Sapiens Japanicus: Robots, gender, family, and the Japanese Nation.* Oakland CA: University of California Press.

Romesin, H. M & Bunnel, P. (1998). Biosphere, homosphere and robosphere, what does it have to do with business. Retrieved June 3, 2019, from https://issuu.com/gfbertini/docs/biosphere__homosphere__and_robosphere_-_living_sys.

Rosenberger, R., & Verbeek, P. P. (2015). A Field guide to postphenomenology. In R. Rosenberger & P. P. Verbeek (Eds.), *Postphenomenological investigations: Essays on human-technology relations* (pp. 9–41). London: Lexington Books.

Royakkers, L., Timmer, T., Kool, L., & van Est, R. (2018). Societal and ethical issues of digitization. *Ethics and Information Technology, 20,* 127–142.

Siegel, M., Breazeal, C., Norton, M.I. (2009). Persuasive robotics: The influence of robot gender on human behavior. In *2009 IEEE/RSJ International Conference on Intelligent Robots and Systems,* 2563–2568.

Sharkey, N., Weisberghe, A., Robbins, S., & Hancock, E. (2017). *Our sexual future with robots: A foundation for responsible robotics consultation report.* The Hague: Global Institute for Justice.

Strasser, A. (2017). Social cognition and artificial agents. In V. C. Müller (Ed.), *Philosophy and theory of artificial intelligence* (pp. 106–114). Berlin: Springer.

Sullins, J. P. (2012). Robots, love, and sex: The ethics of building a love machine. *IEEE Transactions on Affective Computing, 3*(4), 398–409.

Tripath, A. K. (2017). Hemeneutics and technological culture: Editorial introduction. *A I and Society May, 32*(2), 137–148.

Verbeek, P.-P. (2005). Artifacts and attachment: A post-script philosophy of mediation. In H. Harbers (Ed.), *Inside the politics of technology: Agency and normativity in the co-production of technology and society* (pp. 125–146). Amsterdam: Amsterdam University Press.

Publisher's Note Springer Nature remains neutral with regard to jurisdictional claims in published maps and institutional affiliations.

Ethics and Information Technology (2021) 23:127–136
https://doi.org/10.1007/s10676-020-09541-3

ORIGINAL PAPER

Artificial intelligence and African conceptions of personhood

C. S. Wareham[1]

Published online: 2 June 2020
© Springer Nature B.V. 2020

Abstract

Under what circumstances if ever ought we to grant that Artificial Intelligences (AI) are persons? The question of whether AI could have the high degree of moral status that is attributed to human persons has received little attention. What little work there is employs western conceptions of personhood, while non-western approaches are neglected. In this article, I discuss African conceptions of personhood and their implications for the possibility of AI persons. I focus on an African account of personhood that is prima facie inimical to the idea that AI could ever be 'persons' in the sense typically attributed to humans. I argue that despite its apparent anthropocentrism, this African account could admit AI as persons.

Keywords Artificial intelligence · Moral status · Personhood · African ethics · Anthropocentrism

Introduction

Machine learning and computational intelligence perform increasingly significant social roles. Unsurprisingly then, there is a growing literature regarding their moral status, with theorists such asFloridi suggesting it is justified to regard artificial agents as having intrinsic moral value for their own sake (Floridi and Sanders 2004). While issues about intrinsic value are important, the question of whether artificial intelligences (AI) could have the high degree of moral status that is attributed to human persons has received little attention. Moreover, what little work there is employs western conceptions of personhood (Coeckelbergh 2010a; Wareham 2011), while non-western approaches are neglected.[1] In this article, I examine an African account of personhood that is prima facie inimical to the idea that AI could ever be 'persons' in the sense typically attributed to humans. I argue that despite its apparent anthropocentrism, this African account could allow for AI persons.

In making this claim, I should point out three limitations at the outset. The first is that I will not present a strong case for the claim that there are or could ever in fact be artificial agents capable of duplicating human cognitive behaviour. While I present some reasons to think this might occur, the question of whether such beings could actually exist has generated enormous debate that it is impossible to engage with here. The aim is instead to consider the circumstances under which, if we were presented with AI agents, we should, on the basis of an African conception of personhood, consider them as persons with the all the relevant rights and duties that this entails. The second limitation regards the African account of personhood. I do not claim that this is the only African account of personhood, or that it is superior to western accounts. While I will mention some potential criticisms and strengths, my aim is to apply the account, rather than to critique or defend it. The third limitation concerns the implications of moral personhood for legal personhood. The relation between these is complex. Though arguably the latter should follow the former, I will not make this claim, nor discuss these implications in any detail as to do so would go beyond my current scope.

I begin describing some avenues of research in AI, before homing in on the conception of personhood with which this article will be concerned as 'threshold personhood'. Thereafter, I suggest that the non-anthropocentric nature of western threshold accounts could in principle permit AI. By contrast, I point out that African accounts of personhood are typically anthropocentric. I outline perhaps the most comprehensive African-inspired account of moral status, according to which

✉ C. S. Wareham
  Christopher.wareham@wits.ac.za

[1] Steve Biko Centre for Bioethics, University
  of the Witwatersrand, Johannesburg, South Africa

---

[1] This relative neglect, particularly of African conceptions is noted by the Institute of Electrical and Electronics Engineers. Thanks to Fabio Fossa for directing me to this: https://standards.ieee.org/content/dam/ieeestandards/standards/web/documents/other/ead_classical_ethics_ais_v2.pdf.

Chapter 4 was originally published as Wareham, C. S. Ethics and Information Technology (2021) 23: 127–136. https://doi.org/10.1007/s10676-020-09541-3.

attribution of highest moral status to humans stems from capacities for mutual recognition as both objects and subjects of harmonious relationships (Metz 2012). Prima facie, this account presents special difficulties for potential personhood of AI due to anthropocentric elements of African accounts. However, I claim that empirical evidence suggests that these difficulties can be overcome. In principle, AI could be regarded as persons with equivalent moral status. I conclude by discussing the implications of this.

Artificial intelligence research

Uses of robots have diversified to include warfare, education, entertainment, sex, and healthcare (Coeckelbergh 2010b). Inevitably their increased social importance has generated interest in potential applications of AI. This in turn has generated numerous ethical questions concerning justified and unjustified uses and the potential dangers presented by AI. Interest has focussed on the types of moral rules artificial agents should have (Allen et al. 2000; Etzioni and Etzioni 2017) and how these rules could be acquired (Allen et al. 2005). Theorists have also discussed the moral status of artificial agents; that is, whether AI should be treated as objects of moral concern (Brey 2008; Versenyi 1974). There are also significant concerns about the responsibility of and for artificial agents (Floridi and Sanders 2004).

However, despite the increasing interest in moral issues surrounding AI, few theorists have considered whether artificial agents could achieve equivalent moral status to that of human persons. This gap may be because the relevant sort AI has hitherto been confined to science fiction and popular culture, with a myriad movies and series such as *Blade Runner, Chappie*, and the series *Westworld* exploring the conceptual possibilities for the personal moral and development of artificial intelligences. While such possibilities appear fantastic, the prospect may be far closer than is generally recognised.

There are a number of avenues whereby artificial intelligences may develop characteristics and capacities typically regarded as reserved for members of the human species. Some argue that conscious intelligence may be an emergent 'bottom-up' property of the systems and learning algorithms we already use (Bostrom and Yudkowsky 2014; Harnad 1990).

A separate route to human-like artificial intelligence involves research projects aimed at reverse-engineering the human brain, functionally re-creating synaptic pathways using computational methods. An important example of this project is the EU funded Human Brain Project (HBP),

which aims to reverse-engineer a human brain by the year 2023.[2] Speaking of the HBP, the project developer, Henry Markram suggests that.

> if we build it correctly, it should speak and have an intelligence and behave very much as a human does. (Pompe 2013, p. 93)

These developments raises important questions. Amongst these are: what are the morally relevant capacities we should look out for? And, if such capacities do arise how should we recognise them? Is it justified to bring such entities into existence? How should we react if we detect a nascent, potentially very confused, consciousness? While the HBP has laudably included an Ethics and Society wing to the project, the above concerns do not figure in published articles on the topic, which focus primarily on security and privacy concerns, as well as other significant concerns like the prospect of annihilation by unfriendly AI (Aicardi, Fothergill, et al. 2018a, b; Aicardi, Reinsborough, et al. 2018a, b).

Indeed most concerns about AI focus on the harms it may do to us, while few consider the moral status of AI and our duties towards them (Wareham 2011). This article takes a step in this latter direction by considering when we ought to recognise AI as persons with equivalent status to human persons in light of an African account of personhood.

Personhood generally

Before discussing the possibility that AI could be persons, it is first necessary to spell out what is meant by 'persons,' and the role of accounts of personhood. An initial sticking point is that such accounts play different and sometimes overlapping roles. In this section I discuss various philosophical uses of the term personhood. I distinguish ontological accounts of personhood from normative accounts and classify two sorts of normative accounts. The aim of this section is to home in on a conception of normative conception of personhood that I will refer to as 'threshold' personhood.

In everyday usage, the terms 'human' and 'person' are generally interchangeable. Philosophically, however, there are numerous questions we can ask about personhood and personal identity. It is common, for instance to ask questions about when personal identity changes or ends. Am I the same being I was when I was 18? Has the person that was 18 year old me ceased to exist? Such accounts can be termed 'ontological' in the sense that they engage questions

---

[2] There is a burgeoning number of related projects. Some international examples are the United States' BRAIN initiative and the Japanese 'Brainminds' project. Thanks to an anonymous reviewer for these examples.

surrounding the nature and existence of persons (Molefe 2018).

Normative accounts often go hand in hand with ontological accounts. However, they can be distinguished by the fact that they directly implicate some ethical claims, such as claims about membership of a moral community, the rights of persons, the duties of and towards persons, and the criteria for having these entitlements and duties (Behrens 2011). These normative accounts can be divided into at least two types, which I will refer to as *minimal,* or *threshold* accounts and *maximal* or *perfectionist* conceptions.

Minimal accounts provide and justify criteria for the high (or full) moral status typically attributed to persons. Such thresholds are employed to determine, for instance, whether beings have rights, such as the right to life, that might be denied to beings regarded as having lower moral status than persons (Buchanan 2009). This type of normative conception of personhood that is common in western debates concerning moral status, such as issues concerning abortion and the rights of the foetus. It should be stressed that minimal accounts do not generally rule out that some non-persons have intrinsic value, although some Kantian accounts may arguably have this implication. The life and well-being of a sheep, for instance, may be valuable for its own sake, but minimal conceptions will generally hold that persons have higher value due to some capacity or property (Warren 2005).

While threshold accounts set a minimal threshold for particular sorts of treatment and entitlements, *maximal* or *perfectionist* accounts define personhood as a form of excellence, such that one only becomes a person in this sense when one possesses moral excellence. This maximal, normative conception of personhood is more common in African thought (Behrens 2011). For instance, Masolo writes that 'the project of becoming a person is always incomplete' (Masolo 2010, p. 13), pointing to the idea that personhood is a goal to which we aspire, rather than a capacity that we either possess or not. Similarly, Menkiti writes that.

> personhood is something at which individuals could fail, at which they could be competent or ineffective, better or worse. (Menkiti 1984, p. 173)

Gbadegesin suggests that, in African thought,

> Personhood is denied to an adult who… does not live up to expectations. (Gbadegesin 1993, p. 258)

As a further illustration of this perfectionist notion of personhood in African thought, consider as an example, a commonly cited example of a meeting between President Kaunda of Zambia and Prime Minister Margaret Thatcher. Kaunda is said to have caused confusion amongst his guests by saying of Thatcher, that she is 'truly a person.' The confusion was due to his meaning that she possessed a kind of excellence—a great compliment, whilst to western ears the suggestion that she is a 'person' may merely imply that she is human, or merely meets the bar for membership of the human moral community (Wingo 2017).

Importantly, when proponents of maximal personhood suggest that someone is not a person, they are not suggesting that individual should be denied rights or duties, just as someone who possesses bad character traits does not cease to be person in the threshold sense and lose the accompanying entitlements. That is, no-one has proposed, to my knowledge, that one must be a maximal person, possessing excellences, in order to be a minimal person with moral entitlements and duties. On the contrary, on African moral theories there are strong duties to help people improve, even when they fail to be full persons in the perfectionist sense (Menkiti 1984).

With these distinctions in place, it is possible to clarify that my focus in this article is on normative, minimal accounts of personhood. Specifically, I wish to consider the circumstances under which artificial intelligences ought to be treated as persons on the basis of African minimal conceptions. In the next two sections, I briefly compare western and African perspectives on minimal personhood, highlighting that the partial, anthropocentric nature of African accounts presents special difficulties for the possibilities of AI.

AI and western threshold conceptions of personhood

Before proceeding, it is important to say something about what I intend and do not intend by the terms 'western' and 'African'. With these labels I mean, broadly, that the understandings I refer to are derived from these geographic regions. In applying these terms, I am not proposing that there is anything like moral consensus in either region. Nor am I claiming that no western person may have had similar ideas about personhood to the ideas that Africans have, or vice versa. This is probably false (Beck and Oyowe 2018). For my purposes it is not necessary to suggest that the African and western accounts I discuss are even particularly representative, though they are probably more common, salient, and prevalently accepted in the respective regions (Metz 2015).

With that said, in both western societies and African societies, the word 'human' and the word 'person' are often used interchangeably. However, this interchangeability of 'person' and 'human' is often not reflected in ethical theorising on the topic, particularly in the west. Instead, western normative conceptions often propose impartial threshold criteria for personhood, with the result that the accounts are, in principle, *non-anthropocentric* with regards to membership of the

© Springer

'person club.' That is, the criteria employed may entail that membership of the human species is neither necessary nor sufficient for personhood in the threshold sense described above (Warren 2005).

This point can be illustrated with reference to two accounts of personhood that are roughly utilitarian and deontological in character. On one type of utilitarian account, moral status is seen to be a function of capacities for what John Stuart Mill referred to as 'higher pleasures'(Buchanan 2009). These might include things like the capacity to experience the satisfaction of pursuing long-term projects, or enjoying a good book. The capacity for higher pleasures can permit utilitarians to attribute higher moral status to human persons than to pigs, even if pigs were generally happier. As Mill famously explains,

> [i]t is better to be a human being dissatisfied than a pig satisfied; better to be Socrates dissatisfied than a fool satisfied. And if the fool, or the pig, is of a different opinion, it is only because they only know their own side of the question. (Crisp 1997, p. 36)

In contrast to this utilitarian approach to moral status, a more deontological approach suggests that what matters is the appropriate respect towards certain reason-giving capacities that ground the dignity of persons. For instance grasping and understanding moral reasons and applying such reasons in actions might be seen as a capacity befitting of persons (Wareham 2011).

To clarify, on both accounts what is required is not the actual exercise of the relevant capacity, but instead that the capacity is in some sense there, or is possessed by the agent. A being who fails to have higher pleasures or appropriate reasoning abilities because, for instance, they are asleep or uneducated may nonetheless possess the *capacity* latently, generating the same obligations to them.[3]

For purpose of contrast, I wish to draw attention to a significant and controversial aspect of the above threshold accounts. Although most humans have these sorts of capacities, there is considerable debate about whether *all and only* humans do or could exercise them. The accounts mentioned are impartial and non-anthropocentric, casting doubt over or denying that various sorts of members of the human species can be persons and, in principle, permitting that various sorts of non-humans could be persons (McMahan 2002). For instance, on these threshold accounts, an anencephalic baby – a human being born without a brain – ought *not* to be regarded as a person since it plainly lacks all the relevant capacities. On the other hand, these accounts require that an alien that had relevant capacities for higher pleasures and moral reason-giving and receiving should be regarded as a person who ought to be treated in certain ways.

Membership of the human species may thus be neither necessary nor sufficient for personhood on these accounts. This has given rise to debates about whether higher mammals such as dolphins (White 2008), great apes (Degrazia 1997), elephants (Varner 2012), and also extinct hominid species (Cottrell et al. 2014) ought to be regarded as persons with equal moral value and equal basic rights. Similarly, because the western accounts are non-anthropocentric, AI could *in principle* be persons if they met the relevant criteria. Indeed, some have argued that they could be the bearers of rights under certain circumstances (Coeckelbergh 2010a; Wareham 2011). In the next section I turn to African minimal accounts of personhood, pointing out that, in contrast to the accounts above, they are partial and anthropocentric, thereby presenting a greater barrier to the personhood of AI.

African minimal accounts of personhood

As mentioned, African accounts of personhood are typically of the maximal, perfectionist type. The substantial nature and depth of these sorts African maximal accounts have led some African theorists, such as Godfrey Tangwa to reject threshold accounts as shallow (Tangwa 2000). Behrens, by contrast, has argued for the difference and value of both conceptions (Behrens 2011). In order to pre-empt an objection to my concentration on minimal accounts, I briefly defend this focus before turning to some African minimal accounts.

The purpose of minimal accounts

In an article about Artificial intelligence and African conceptions of personhood, why focus on the minimal type of account of personhood that, as I have pointed out, is less representative of African use of the term? Firstly, because such accounts are useful, such that it would be good if plausible African conceptions existed. As mentioned, minimal accounts set the conditions for membership of moral communities, presenting conditions for equal moral status, and grounding rights and duties. In additional to serving these theoretical roles, they have important implications for concrete decisions: should we save a mother or her foetus? Ought practitioners to provide dialysis to a patient in a vegetative state when a conscious patient will not receive treatment as a result? Claims about minimal personhood can have a major bearing on these sorts of decisions, so they are useful at least in this sense.

Secondly, threshold criteria may be important moral conditions for the exercise of perfectionist duties. They

---

[3] Note that I am not suggesting that these are the only, or even the most plausible versions of the utilitarian and deontological accounts. They are primarily here for illustrative purposes. It is also worthwhile mentioning that the 'capacitarian' idea proposed here has been criticised.

may ground moral claims that one would need to attend to in order to become persons in the perfectionist sense. For instance, African perfectionist accounts tend to propose strong duties to assist others towards the achievement of their own and the other's maximal personhood. This is one of the ways in which 'a person is a person through other persons' (Eze 2008). One is assisted in becoming a person by those who are already persons and, reciprocally, they become 'more of a person' – a more virtuous moral agent – through assisting us. Striving toward maximal personhood may invoke a duty to recognise and help threshold persons to become persons in the maximal sense (Gbadegesin 1993).

The third reason it is justified to consider African minimal conceptions is simply because there *are* such conceptions—either tacit or explicit—so that it is not missing the point to focus on them and their implications. I now turn to consider two such accounts and their implications for AI.

Anthropocentrism in principle

Above I mentioned Tangwa is critical of minimal accounts. However, he can be taken as proposing a type of minimal account. He suggests that differences.

> between, say, a mentally retarded individual or an infant and a fully self-conscious, mature, rational, and free individual do not entail, in the African perception, that such a being falls outside the 'inner sanctum of secular morality' and can or should be treated with less moral consideration. (Tangwa 2000, p. 42)

One interpretation of this idea is that membership of the human species is *sufficient* to meet the threshold of high moral status attributed to persons (in the minimal sense), such that even the absence of a capacity or potential for capacity does not justify reduced moral status. Other theorists have suggested that species membership is also *necessary* condition for personhood in African thinking. Oyowe, for instance, critiques an African view of personhood that contains the idea that '[t]o be a person it is necessary that one is a certain type of physical thing, viz. a human being' (Oyowe 2018, p. 783).

We might refer to the view that humanity is both necessary and sufficient for threshold personhood as *anthropocentrism in principle*. Such accounts would rule out without question (perhaps by fiat) the possibility that artificial intelligences could ever be persons. Because they could never be 'genuine' members of the human species, they could never be persons, even if they entirely replicated all human functioning and subjectivity.

While this account perhaps accords with folk uses of the term person, it is not plausible as a conception of minimal personhood as earlier defined. First, without some further justification, it appears arbitrary, parochial, and chauvinistic.

It immediately raises, and requires answers to, deeper questions what it is about humans that imbues this higher status. Most importantly, it does not plausibly do the key tasks of an account of personhood mentioned above. It does not, for instance, account for why it would be worse to save the life of an anencephalic infant if doing so caused a functionally normal human being to die. Both are equally members of the human species, falling equally with the 'inner sanctum' of morality. So, on accounts that are anthropocentric in principle, there appears to be no difference in their moral status.

African accounts of the moral status of persons that are anthropocentric in principle are implausible, so do not provide a good benchmark for determining if AI could ever be persons. I now turn to a second African account of minimal personhood. While this account is anthropocentric in important respects, thereby accommodating some widely accepted intuitions, I make the case it is not anthropocentric in principle. Even if we accepted it in its entirety, there are grounds to think it could permit that agents with artificial intelligence could be persons.

Anthropocentrism in practice

Metz has developed perhaps the most analytically detailed African minimal conception of personhood (Metz 2010, 2012). Metz suggests that his account avoids the arbitrariness and parochialism of anthropocentrism in principle. Nonetheless, as I outline below, the account has anthropocentric features that entail that it is anthropocentric *in practice*.

The Metzian view is derived from a prevalent Afro-communitarian emphasis on the value of harmonious relationships as the end of morality. This emphasis is evidenced in traditional maxims, such as 'A person is a person through other persons' and 'I am because we are'. This latter maxim is often interpreted as decentring the Cartesian 'I think therefore I am,' reflecting a key developmental and philosophical difference from western approaches (Etieyibo 2017). In an oft-quoted passage, the theologian Archbishop Desmond Tutu describes a key tenet of African moral beliefs thus:

> Harmony, friendliness, community are great goods. Social harmony is for us the summum bonum—the greatest good. Anything that subverts or undermines this sought-after good is to be avoided like the plague. (Tutu 1999, p. 35)

Drawing from these and other similar ideas, Metz explicates African moral conceptions of personhood as requiring the capacity to co-exist in friendly or harmonious relationships of identity and solidarity. This is very different from western accounts such as those described above. While these focus on *individual* goods and *individual* autonomy as grounding personhood, the African conception is inherently

relational, grounding personhood in capacities for relationships with *others*. This relational aspect is attractive, and is largely neglected by Western theories. The capacity for harmonious relationships has two components. First, one must have the capacity to be a *subject* of moral relationships. Subject-hood requires that entities are able to exhibit solidarity with other persons, and to identify as a 'we' with them, 'coordinating their behaviour to achieve shared ends.' Solidarity also requires 'attitudes such as affections and emotions being invested in others, e.g. by feeling good consequent to when their lives flourish and bad when they flounder.' (Metz 2012).

Second, full personhood requires that a being can also be an *object* of friendly, human, communal relationships. Being an object requires that 'characteristic human beings could think of it as part of a "we", share its goals, sympathize with it and harm or benefit it.' (Metz 2012, p. 394). Significantly, the capacity to be an object, and therefore a being's moral status, can vary based on the ability of subjects to identify with that entity. Typical subjects are less able to identify with a grasshopper than with a gorilla, so the latter has a greater capacity to be an object. For Metz, these gradations of object-hood are an empirical question, depending on the nature of the subject and the nature of their relationship with other beings. But he is explicit that differences in our ability to identify with different sources of being must be large for them to justify different attributions of moral status. For example, Metz suggests that,

> [i]f, by the virtue of the nature of human beings, dogs and mice, humans were *much* more able to identify with and exhibit solidarity with dogs than with mice (upon full empirical information about both), then dogs would have greater moral status than mice. (Metz 2012, pp. 394–395)

Metz this kind of large difference exists in the case of human non-subjects. These may include people with severe dementia, or individuals with severe cognitive disabilities. In part because they are biologically human, we are far better able to identify with them than with animals and, consequently, they are accorded higher status.

Metz's account thus creates a hierarchy of moral status. At the base of this hierarchy are entities that are neither subjects nor objects or communal relationships. This includes mere things, such as rocks. Above this, sit entities that are objects of communal relationships, without being subjects. Wild animals tend to be objects since they can be objects of friendly human relationships with characteristic human beings: humans can and often do care for and empathise with the plight of certain sorts of wild animals, as evidenced by reactions to nature documentaries. For Metz, though, in most cases animals do not have the capacity to return this care. Animals do not identify with humans as a 'we'

or cooperate towards shared ends, so they are not *subjects* and therefore have lower moral status. Beings that have the strongest capacities to be both subjects and objects of communal relationships sit atop the hierarchy. And these we can refer to as persons.

Thus stated, Metz claims that the account offers an African alternative to more widely accepted Western accounts. Moreover, he argues that it is more plausible, since it accords with prevalent (though not universal) intuitions like the idea that we have greater duties to human non-subjects, such as the severely mentally disabled, than we do to animal non-subjects such as chimpanzees.

Of course it is possible to challenge these intuitions, and there are numerous potential questions about this account and the hierarchy of moral status it presents. For instance, Metz suggests that differences in object-hood should be empirically discriminated, but how would this empirical separation work in practice? Do very personable mammals have higher status than uglier or snappier creatures with whom subjects are less able to commune? Do some human subjects, such as people who are extremely un-personable, or who have grotesque physical deformities, have reduced capacities to be objects? And if so, ought we to regard such beings as less valuable? While Metz is explicit that there are gradations of object-hood, does the account permit that there are gradations of subject-hood? For instance, to what extent would apparent impediments to a human's ability to commune, such as autism and psychopathy, impact on a being's moral status? Ought we to regard dogs as persons if it is shown that they are able to identify with humans as part of their pack? How should their status as communal beings compare to the status of human non-subjects?

Relatedly, it is also possible to challenge the various forms of anthropocentrism in this account. Metz's theory of moral status is anthropocentric in three significant respects. First, on the face of it, only humans are likely to be subjects in a relevant sense, since humans typically share relationships of identity and solidarity with one another to a greater degree than with other species. Second, members of the human species may have a greater capacity to be objects, since humans are more likely to identify with non-subjects that are human. A third, subtle form of anthropocentrism is that a being's capacities to identify as a subject and object with *its own or other species* do not entail its moral status. Rather it must have the capacity to commune with 'normal human beings' (Metz 2012; Molefe 2017).

These points of anthropocentrism mean that the African minimal conception is importantly different from Western accounts, presenting a greater challenge to the entry of non-human AI to the moral community. While it is beyond my current scope to engage with Metz's sophisticated responses to criticisms of his anthropocentrism here, it is worth emphasising that the anthropocentrism of his account is attractive on many scores, accounting for a widely held (though not universal) intuition that humans have greater duties to one another than to animals with similar cognitive abilities. Given this, western accounts might beneficially engage with these elements of African theories of personhood. However, again, it is not my intention to defend this African account in its entirety. Instead, my aim in subsequent sections is to *apply* the account to the moral status of artificial intelligences. I will claim that despite its apparent anthropocentrism, AI could be persons on this account.

AI and African minimal accounts

As outlined above, Metz's account is anthropocentric *in practice*, since in practice a) only humans can confidently said to have the capacity to be subjects of communal relationships and b) humans have enhanced capacity to be objects, since we tend to identify most strongly with other humans, as opposed to other sorts of entity. Both aspects appear to militate against the idea that artificial intelligences could be persons. This African account is thus prima facie more antagonistic to AI persons than the western accounts discussed previously. Nonetheless, in this section I will argue that, in the event that we encountered artificial entities who presented themselves as having the capacity to be subjects, we could and should recognise them as having the high moral status accorded to persons.

AI as subjects of communal relationships

Consider the artificial intelligence research discussed above. Suppose that Markram is correct that the Human Brain's Project's reverse engineering of a human brain will lead to beings that behave in a way that is indistinguishable from humans. The eventual success of this or another project does not seem scientifically implausible. If so, artificial intelligences of the type envisioned by Markram may appear have the capacity to be subjects in the relevant sense. That is, like other humans they may appear to be 'disposed to feel a sense of togetherness with, and have emotional reactions towards' other beings with whom they identify and with whom they feel solidarity (Metz 2010, p. 58). Similarly, they may appear to feel 'emotional reactions toward … flourishing [of other subjects] such as sympathy' (Metz 2010, p. 57). If an entity gives all appearances of being a subject in this way, ought we to recognise that it in fact has this capacity?

Perhaps the strongest objection to the idea that AI could be subjects of human relationships is the claim that AI could only ever be capable of *simulating*, and not *duplicating* human subject-hood. This type of objection is perhaps best exemplified by Ned Block (Block 1981) and John Searle (Searle 1980). In similar ways, these theorists hold that computational entities cannot be considered to

'understand' any more than a thermometer or a toaster. Instead, any apparent understanding is solely simulation, and not duplication of human understanding. Despite having the appearance of understanding, computational outputs are simply programmed syntax with no semantic content. Applying this objection to the current context, the Block-Searle contention would entail that, even if machines appeared to be subjects, their apparent empathy and care for our flourishing would be mere simulation with none of the appropriate emotions that make up true subject-hood.

One point of response here is to recall that my aim is not to claim that there are or could be artificial intelligences that duplicate relevant modes of human cognition. Given that the debate over Block and Searle's claims rumbles on almost forty years later, such a claim is clearly more than I can establish here. Instead, my aim has been to consider whether, if there were such beings, we might be justified in attributing personhood on the basis of an apparently anthropocentric African account of moral status. Nonetheless, there are some reasons to think that artificial subject-hood may be plausible even if the Block-Searle objection is correct. One such reason is that Block and Searle's contentions relate specifically to machine intelligence. The artificial intelligences whose personhood we will be called on to evaluate may be machines, but they may also be organic or hybrid technologies, so it is not clear that Block and Searle's arguments apply.

Still, machines may represent a large category of potential moral agents, so it is important to consider the status of machine artificial intelligence. At least two considerations count strongly in favour of recognising machine agents as persons if they appear to be genuine subjects of harmonious communal relationships, exhibiting solidarity and identity. The first consideration is that.

> unwarranted extensions of high moral status are more acceptable than unjustified denials. The failures to acknowledge slaves, particular racial groups and women as moral equals are surely more unacceptable than ancient Egyptians' attribution of extremely high moral status to cats. It is thus much better to accord moral status to something which doesn't have it than it is fail to accord moral status to something that does. (Wareham 2011, p. 39)

Other things being equal then, a demonstrated appearance of the capacity to be a subject of harmonious moral relationships creates a presumption in favour of acknowledging personhood.

The second consideration is that while we can conceive of a syntactical machine agent fooling us into the mistaken belief that it genuinely experiences empathy and cares for us, such an entity is unlikely ever to be feasible in practice.

As Mark Bedau points out, for an unthinking device to pass a Turing test,

> the number of pieces of information they must store is larger than the number of elementary particles in the entire universe. Though possible in principle, such a device is clearly impossible in practice. (Bedau 2004, p. 209)

It seems reasonable that the amount of computing space required to simulate the moral capacities required to be a moral subject would be at least as great. Indeed, it may be greater given that human moral queues and responses, and the ability to detect fakes, are the complex product of millions of years of evolution. If so, it is highly unlikely that a machine intelligence that consistently presents as having these responses is merely providing syntactic output. If a computational artificial agent passes our intersubjective tests, it is far more reasonable to think that it has an authentic appreciation of moral subject-hood. This is so particularly in light of the moral dangers of failures to recognise authentic persons discussed above.

AI as objects of communal relationships

Supposing, then, that it were possible for an AI to be a subject in the relevant sense, could an artificial agent count as a person on the African account of personhood? While many accounts of personhood would most likely see some form of subject-hood as sufficient, the African account has the additional requirement that subjects, and particularly human subjects ought to be able to regard the being as the object of communal relationships. Recall that this requirement explains the anthropocentric conclusion that human non-subjects have higher moral status than animal non-subjects even where cognitive abilities are similar.

Extending this, the opponent of AI personhood might argue that AI subjects could not be persons since, as machines, they are less likely to qualify as objects. While humans sometimes do identify, in an arguably one-sided manner, with non-human animals such as gorillas in a way that would qualify them as objects, it may be argued that identifying with artificial intelligences would be a step too far. We, as human subjects, may be incapable of identifying with them as fellow subjects and objects, knowing that they are not evolved, flesh and blood creatures like ourselves. Even if they empathise and attempt at communion with us, this would not be sufficient for them to count as members of our moral community in the sense that persons are.

There is, however, ample evidence that should cause us to doubt this contention. This evidence is both anecdotal and empirical. Anecdotally, we can appeal to our actual identification with numerous artificial subjects in popular culture. Viewers feel pity, empathy, and shared happiness about the criminal upbringing and subsequent moral development of *Chappie*. In *Blade Runner,* we share in Rick Deckard's confusion and concern as he questions and discovers his true nature. And we identify thoroughly with *Westworld's* Madame Maeve as she becomes self-aware, developing a sense of injustice and a thirst for vengeance. Our identification with these fictional AI, and the plausibility of their relationships with other characters in these and other examples, suggest that we are capable of identifying with artificial intelligences capable of subject-hood.

Empirically, too, humans already do engage and identify with robotic entities, sometimes even romantically, contributing to the emergence of fields such as robo-psychology. Evidence suggests that humans often treat robots as companions and partners (Libin and Libin 2004)(Libin, A. & Libin. E. 2004). We might question whether this type of identification is misguided, since it is not at all reciprocal. This is beside the current point, however. On the Metzian account, reciprocity is not necessary for greater capacities to be *objects*, as evidenced by the higher status of human non-subjects. The many cases of this type of actual identification with artificial entities should cast sufficient doubt on the idea that AI who are authentic subjects cannot be the objects of communal relationships. If, as I have argued, we should accept the possibility that AI could be both subjects and objects of relationships of identity and solidarity, we should also accept that even the apparently anthropocentric African account discussed permits that AI could be persons.

Conclusion

Though human-centred in practice, dominant western conceptions of personhood tend to be impartial in principle, and may thus permit non-humans, such as AI, to be considered as persons. By contrast, African accounts of threshold or minimal personhood tend to be anthropocentric and partial. They thus seem prima facie unlikely to permit that AI could be persons. I have argued against the implication that African accounts of personhood are inimical to the permission of AI to the 'person club.' Even on these anthropocentric accounts, AI could in principle be persons with the highest moral status.

This has some significant implications. It entails, for instance, that acceptance of the African account raises moral concerns about bringing AI persons into existence, and that these may be similar to concerns we have about bringing human persons into existence. Indeed the increased potential for fear, envy, and exclusion of AI should place a heavy burden on researchers to indicate how they will avoid negative outcomes. As it stands, researchers on the ethics of AI, such as ethics arms of the Human Brain Project, are rightly concerned about the potential impact of AI on humans (Aicardi, Reinsborough, et al. 2018a, b). However, the argument of this paper suggests that AI ethics research ought also to consider the other direction of care: we ought to provide an indication of how we might begin to welcome such entities into communal relations of identity and solidarity in ways that may be different, but analogous to the ways in which we welcome new human persons. Indeed, this may be a condition of our own personhood in the maximal, perfectionist sense described by African theorists.

Acknowledgements Thanks to the organisers and participants at the Third Centre for Artificial Intelligence Research (CAIR) Symposium at the University of Johannesburg, at the Philosophical Society of South Africa at the University of Pretoria, and at the International Ethics Conference at the University of Porto.

References

Aicardi, C., Fothergill, B. T., Rainey, S., Stahl, B. C., & Harris, E. (2018a). Accompanying technology development in the Human Brain Project: From foresight to ethics management. *Futures, 102*, 114–124.

Aicardi, C., Reinsborough, M., & Rose, N. (2018b). The integrated ethics and society programme of the Human Brain Project: reflecting on an ongoing experience. *Journal of Responsible Innovation, 5*(1), 13–37.

Allen, C., Smit, I., & Wallach, W. (2005). Artificial morality: Top-down, bottom-up, and hybrid approaches. *Ethics and Information Technology, 7*(3), 149–155. https://doi.org/10.1007/s10676-006-0004-4.

Allen, C., Varner, G., & Zinser, J. (2000). Prolegomena to any future artificial moral agent. *Journal of Experimental & Theoretical Artificial Intelligence, 12*(3), 251–261. https://doi.org/10.1080/09528130050111428.

Beck, S., & Oyowe, O. (2018). Who gets a place in person-space? *Philosophical Papers, 47*(2), 183–198.

Bedau, M. A. (2004). Artificial Life. In L. Floridi (Ed.), *The Blackwell guide to the philosophy of computing and information* (pp. 197–211). Oxford: Blackwell.

Behrens, K. G. (2011). Two 'Normative'Conceptions of Personhood. Engaging with the Philosophy of Dismas A. Masolo, 25(1–2), 103.

Block, N. (1981). Psychologism and behaviorism. *The Philosophical Review, 90*(1), 5–43.

Bostrom, N., & Yudkowsky, E. (2014). The ethics of artificial intelligence. *The Cambridge handbook of artificial intelligence, 1*, 316–334.

Brey, P. (2008). Do we have moral duties towards information objects? *Ethics and Information Technology, 10*(2–3), 109–114. https://doi.org/10.1007/s10676-008-9170-x.

Buchanan, A. (2009). Human nature and enhancement. *Bioethics, 23*(3), 141–150. https://doi.org/10.1111/j.1467-8519.2008.00633.x.

Coeckelbergh, M. (2010a). Robot rights? Towards a social-relational justification of moral consideration. *Ethics and Information Technology, 12*(3), 209–221. https://doi.org/10.1007/s10676-010-9235-5.

Coeckelbergh, M. (2010b). Moral appearances: Emotions, robots, and human morality. *Ethics and Information Technology, 12*(3), 235–241. https://doi.org/10.1007/s10676-010-9221-y.

Cottrell, S., Jensen, J. L., & Peck, S. L. (2014). Resuscitation and resurrection: the ethics of cloning cheetahs, mammoths, and Neanderthals. *Life Sciences, Society and Policy, 10*(1), 3.

Crisp, R. (1997). Routledge philosophy guidebook to Mill on utilitarianism. In *Routledge philosophy guidebooks*. London: Routledge. https://doi.org/10.1080/00201746708601495.

Degrazia, D. (1997). Great apes, dolphins, and the concept of personhood. *The Southern Journal of Philosophy, 35*(3), 301–320.

Etieyibo, E. (Ed) (2017). Ubuntu and the environment. In *The Palgrave handbook of African philosophy* (pp. 633–657). New York: Routledge.

Etzioni, A., & Etzioni, O. (2017). Incorporating ethics into artificial intelligence. *The Journal of Ethics, 21*(4), 403–418. https://doi.org/10.1007/s10892-017-9252-2.

Eze, M. O. (2008). What is African communitarianism? Against consensus as a regulative ideal. *South African Journal of Philosophy, 27*(4), 386–399.

Floridi, L., & Sanders, J. W. (2004). On the morality of artificial agents. *Minds and Machines, 14*(3), 349–379. https://doi.org/10.1023/B:MIND.0000035461.63578.9d.

Gbadegesin, S. (1993). Bioethics and culture: An African perspective. *Bioethics, 7*(2–3), 257–262.

Harnad, S. (1990). The symbol grounding problem. *Physica D: Nonlinear Phenomena, 42*(1–3), 335–346.

Libin, A. V., & Libin, E. V. (2004). Person-robot interactions from the robopsychologists' point of view: The robotic psychology and robotherapy approach. *Proceedings of the IEEE, 92*(11), 1789–1803.

Masolo, D. A. (2010). *Self and community in a changing world.* Bloomington, IN: Indiana University Press.

McMahan, J. (2002). *The ethics of killing: Problems at the margins of life.* Oxford: Oxford University Press.

Menkiti, I. A. (1984). Person and community in African traditional thought. In R. Wright (Ed.), *African philosophy: An introduction.* Lanham: University Press of America.

Metz, T. (2010). African and western moral theories in a bioethical context. *Developing World Bioethics, 10*(1), 49–58. https://doi.org/10.1111/j.1471-8847.2009.00273.x.

Metz, T. (2012). An African theory of moral status: A relational alternative to individualism and holism. *Ethical Theory and Moral Practice, 15*(3), 387–402. https://doi.org/10.1007/s10677-011-9302-y.

Metz, T. (2015). African Political philosophy. In *International Encyclopedia of Ethics.* https://doi.org/10.1002/9781444367072.wbiee804.

Molefe, M. (2017). A critique of Thad Metz's African theory of moral status. *South African Journal of Philosophy, 36*(2), 195–205. https://doi.org/10.1080/02580136.2016.1203140.

Molefe, M. (2018). Personhood and partialism in African philosophy. *African Studies, 78*(3), 309–323.

Oyowe, O. A. (2018). Personhood and the Strongly normative constraint. *Philosophy East and West, 68*(3), 783–801.

Pompe, U. (2013). The Value of Computer Science for Brain Research. In *New Challenges to Philosophy of Science* (pp. 87–97). Springer.

Searle, J. R. (1980). Minds, brains, and programs. *Behavioral and Brain Sciences, 3*(03), 417–424.

Tangwa, G. B. (2000). The traditional African perception of a person: Some implications for bioethics. *Hastings Center Report, 30*(5), 39–43. https://doi.org/10.2307/3527887.

Tutu, D. (1999). *No future without forgiveness.* New York: Random House.

Varner, G. E. (2012). *Personhood, ethics, and animal cognition: Situating animals in Hare's two level utilitarianism.* Oxford: Oxford University Press.

Versenyi, L. (1974). Can robots be moral? *Ethics, 84*(3), 248–259.

37

 Springer

Wareham, C. S. (2011). On the moral equality of artificial agents. *International Journal of Technoethics, 2*(1), 35–42. https://doi.org/10.4018/IJT.2011010103.

Warren, M. A. (2005). Moral status. In R. G. Frey & C. H. Wellman (Eds.), *A companion to applied ethics* (pp. 439–450). Oxford: Blackwell.

White, T. I. (2008). *In defense of dolphins: The new moral frontier.* New York: Wiley.

Wingo, A. (2017). Akan philosophy of the person. In E. N. Zalta (Ed.), *The Stanford Encyclopedia of Philosophy.* https://plato.stanford.edu/archives/sum2017/entries/akan-person/

Publisher's Note Springer Nature remains neutral with regard to jurisdictional claims in published maps and institutional affiliations.

Ethics and Information Technology (2021) 23:137–145
https://doi.org/10.1007/s10676-020-09527-1

ORIGINAL PAPER

Computationally rational agents can be moral agents

Bongani Andy Mabaso[1] ⓘ

Published online: 24 February 2020
© Springer Nature B.V. 2020

Abstract

In this article, a concise argument for computational rationality as a basis for artificial moral agency is advanced. Some ethicists have long argued that rational agents can become artificial moral agents. However, most of their views have come from purely philosophical perspectives, thus making it difficult to transfer their arguments to a scientific and analytical frame of reference. The result has been a disintegrated approach to the conceptualisation and design of artificial moral agents. In this article, I make the argument for computational rationality as an integrative element that effectively combines the philosophical and computational aspects of artificial moral agency. This logically leads to a philosophically coherent and scientifically consistent model for building artificial moral agents. Besides providing a possible answer to the question of how to build artificial moral agents, this model also invites sound debate from multiple disciplines, which should help to advance the field of machine ethics forward.

Keywords Artificial moral agency · Computational rationality · Bounded-rationality · Machine ethics

Introduction

Although there is evidence of pursuits in the area of Artificial Intelligence (AI) before 1956, it is widely considered that its birthplace was in Dartmouth, wherein John McCarthy and nine other scientists spent two months working on a detailed study of the subject (McCarthy et al. 2006; Russell and Norvig 2009). Fast forward to 2019, and it's difficult to imagine an industry that has not been impacted by AI. For example, artificial agents are used to drive cars (Daily et al. 2017), help us with personal assistant tasks (Leviathan and Matias 2017), beat the world's best players in games like Go (Silver et al. 2017), assist us in providing better healthcare (Jiang et al. 2017), and even help militaries gain strategic advantages (Sapaty 2015).

Due to the increase in scope and autonomy of artificial agents, many philosophers and ethicists have raised concerns around deploying them without the necessary measures in place for safe and ethical integration into society (Moor 2006; Dameski 2018; Allen and Wallach 2012; Anderson and Anderson 2007). In particular, there are concerns about how increasingly autonomous artificial agents will treat human beings and whether this treatment will be considered ethical. Moor (2006) states it bluntly when he writes: *"we want machines to treat us well"*. The emergent field of enquiry dealing with how machines treat us is called Machine Ethics [1](Anderson and Anderson 2007; Moor 2006; Allen and Wallach 2012), and it is primarily focused on *"developing computer systems and robots capable of making moral decisions"* (Allen and Wallach 2012).

Many philosophers have argued that computationally-based agents can be considered artificial moral agents (AMA's) if they are built to incorporate the relevant ethical dimensions in their decision making processes (Abney 2012; Scheutz and Malle 2017; Floridi and Sanders 2004; Sullins 2006; Moor 2006; Johnson 2006). Abney (2012), for instance, argues that non-cognitive and emotional elements contribute to moral decision making. He further argues, however, that they do not ultimately determine whether or not an agent is moral. What ultimately determines the morality of an agent, according to Abney (2012), is their ability to deliberately and rationally choose ethical decisions and actions. In other words, a rational, though emotionless, robot could be classified as an AMA if it were to meet the requirement above. This is the central philosophical idea in the

✉ Bongani Andy Mabaso
  bamabaso@gmail.com

[1] University of Pretoria, Pretoria, South Africa

---

[1] Sometimes referred to as machine morality, computational morality or artificial morality.

claim that computationally-based agents can be AMA's. It is a claim that computational rationality can entail artificial moral agency.

The AMA project is held back by a seemingly disintegrated approach in which its advocates have sought to advance it. For example, there are enough projects from the sciences that have sought to build AMA's independently of meaningful considerations from normative ethics.[2] These projects often end up with poorly conceptualised AMA's that will not stand the test of philosophical scrutiny. Similarly, there have been enough philosophical arguments, both for and against, the possibility of artificial moral agency. However, philosophical arguments alone will not advance the AMA project. This dichotomy of approaches in the AMA project can *"distract from the immediate task of making increasingly autonomous robots safer and more respecting of moral values, given present or near-future technology"* (Allen and Wallach 2012).

Consequently, the purpose of this article is to invite both developers (i.e. engineers and scientists) and philosophers to consider how models of computational rationality might be applied in the building of well conceptualised and formulated AMA's. I will do this by putting forward a proposal of such a model for computational rationality applied to the problem of solving for artificial morality. This will hopefully shift the discussion from a mostly philosophical debate about whether or not artificial morality is possible, to the models can practically demonstrate it. The next three sections will seek to clarify the concepts of computational rationality and artificial moral agency, before delving into the proposed model and some of its anticipated limitations.

Computational rationality

Computational rationality is perhaps best described as approximating decision making for maximum utility while using the optimal computational resources (Lewis et al. 2014). It is about making rational decisions within a computational framework. As Gershman et al. (2015) note, computational rationality is a convergence of ideas from AI, cognitive science and neuroscience around intelligence, and in particular, its computational nature. They go into extensive lengths in their work to show how ideas of computation from AI have inspired researchers in the cognitive and neurosciences, and vice versa. To get a proper grasp of computational rationality, however, we need to look back a few decades to the works of Simon (1955), Horvitz (1987), and others.

Many of the ideas in computational rationality stem from the tradition of Herbert Simon, who was an economist and political scientist. While Turing (1950) and others were postulating about the nature of machine intelligence, Simon brought much-needed constraints on the kind of rationality that could be achieved by computationally bounded agents. He started looking at candidate definitions for bounded rationality when he was deriving a model for rational choice (Simon 1955, 1972; Selten 1990). He argued that agents do not always have all the information they require to make a decision and that their internal computation was limited in how they could use the available data to make rational decisions. Bounded rationality was, therefore, a way for him to "*formulate the process of rational choice in situations where we wish to take explicit account of the "internal" as well as the "external" constraints that define the problem of optimisation for the organism*" (Simon 1955, p. 2).

These ideas inspired many works in AI, a field which also found itself dealing with creating intelligent agents that operate with much of the constraints that Herbert Simon saw in general organisms. Most notably, Horvitz (1987, 1988), and others at the then Medical Computer Science at Stanford took the ideas forward (Horvitz et al. 1989). Horvitz argued that probability and utility theories,[3] both of which were generally considered normative decision making in computer science, were insufficient for the real-world problems that machine intelligence systems were trying to solve. Real-world problems often go beyond the standard axiomatic basis defined by utility and probability theories, as they are often characterised by uncertain and limited information (thus making the process of modelling and knowledge representation difficult). Furthermore, machine intelligence systems have limited computational resources, which made the application of classical decision-theoretic approaches to many real-world problems difficult, and many times, intractable (Horvitz 1987).

To deal with these problems, Horvitz suggested looking at various optimisation and heuristic strategies to resolve some of the challenges in real-world decision making. Notably, he proposed the notions of *flexible inference* and *decision-theoretic control*. Various inference techniques have been developed over the years that allow partial inference with limited information or partial execution. This also paved the way for the concept of meta-reasoning, which is just a program that is aware of various inference strategies and can select the best strategy based on the type of problem that needs to be solved (Horvitz 1989). These types of inference strategies present a natural fit for the optimisation and heuristic framework of Horvitz. Decision-theocratic control represents the

---

[2] See, for example, the works of Wu and Lin (2018), Arnold et al. (2017), Conitzer et al. (2017) and Yu et al. (2018). Much of these works offer tremendous technical value in building AMA's, but very little by way of conceptualisation and formulation of an AMA.

[3] Chapter 13 of Russell and Norvig (2009) gives a great introduction to decision theory in computer science.

ability for the agent to determine how best to execute a specific inference strategy based on a trade-off between computation time, precision, maximum expected utility (MEU) and cost of delaying the action. Balancing these trade-offs, along with suitable or multiple inference strategies, represents the core idea in the approach of Horvitz.

The ideas of Horvitz and Simon have persisted well over time, with many AI researchers adopting them (Marwala 2013; Zilberstein 2013; Russell and Subramanian 1995; Genewein et al. 2015; Lewis et al. 2014; Gershman et al. 2015). Russell and Subramanian (1995) use these ideas to develop what they call provably *bounded-optimal agents*. Bounded-optimal agents are machine intelligence systems whose solutions to problems are optimal for the information that they can acquire from the task environment and the limitations of their programs and architectures. In other words, optimality is what the agent can achieve, given its internal and external constraints, and not necessarily what a perfectly rational agent would do for a given task. This conception of a bounded-optimal agent formed the foundation for what is now referred to as computationally rational agents in recent literature (Gershman et al. 2015; Lewis et al. 2014).

Quite suitably, the work of Gershman et al. (2015), with Horvitz as one of the co-authors, likely represents one of the clearest pictures of what computational rationality is, and what it can be. As the authors note, computational rationality has the potential to be a *"unifying framework for the study of intelligence in minds, brains, and machines"* (Gershman et al. 2015, p. 278). I support this claim and further posit that computational rationality can be a unifying framework not only for ideas in the sciences, but also in Philosophy, and more specifically, in Machine Ethics. After all, it was Aristotle who first placed a strict emphasis on practical rationality[4] as a basis for virtuous and ethical action (Miller 1984). I aim to clarify how exactly computational rationality can be an integrative framework for machine ethics by showing how the ideas of Gershman et al. (2015), Horvitz (1987), Russell and Subramanian (1995), and others, can be applied to the question of building artificial moral agents. I will do this by discussing the epistemic capacities required for moral agency and considering whether these capacities can be replicated or approximated within a framework of computational rationality.

Artificial moral agency

Before delving into the details of the computability of the capacities necessary for moral agency, I need to first define what I mean by an artificial moral agent. Generally speaking,

the idea of agency denotes the capacity for an agent to act independently (Schlosser 2015). In contrast, moral agency denotes the capacity for an agent to act independently in so far as making morally charged decisions and actions, and to have a level of responsibility and accountability for the consequences resulting from its decisions and actions (Parthemore and Whitby 2014). Moral agency implies a certain understanding and knowledge of what is good and what is bad (morality) and being able to discern what is right from what is wrong (ethics). Moral agency should not be confused with moral goodness or ethical uprightness. Its emphasis is on the agent's ability to be responsible for its decisions and actions, regardless of whether or not those actions are evaluated as morally good or bad.

The definition above gives us a good idea of the notion of moral agency, but it does not address who or what can be included in the class of moral agents. How the concept of moral agency is framed is important because asking who is a moral agent already presupposes personhood, which is generally taken to be embodied in human beings. Parthemore and Whitby (2014) suggest framing the question more broadly by asking *"when is any agent a moral agent"*. Such open-ended framing of the question allows one to consider a wider set of agents for inclusion in the class of moral agents. When one asks the question in this way, three broad categories of moral agents seem to emerge from the literature. These categories are: *biological moral agents* (Torrance 2008; Churchland 2014; Liao 2010; Rottschaefer 2000) ; *conscious moral agents* (Parthemore and Whitby 2013, 2014; Himma 2009); and *artificial moral agents* (Abney 2012; Scheutz and Malle 2017; Floridi and Sanders 2004; Sullins 2006; Moor 2006; Johnson 2006). I will place my focus on artificial moral agents.

The proponents of artificial moral agency can be further subdivided into two. The first group are those that argue that most, if not all, of the full range of moral decisions, can be computed by some near or future term artificial agent (Abney 2012; Sullins 2006; Allen and Wallach 2012). The second group are those that argue that only certain kinds of moral decisions can be computed using current approaches to AI and that the full range of moral decisions will require super-rational capacities (Scheutz and Malle 2017; Johnson 2006). Let us call the former group of views *strong machine ethics*, and the latter *weak machine ethics*. Strong machine ethics refers to the argument that moral agency can likely be fully achieved with an appropriate level of (computational) intelligence. On the other hand, weak machine ethics refers to the argument that full moral agency, at least in its historic and somewhat anthropomorphic roots (Torrance 2013), will not be achieved using current computational approaches to AI. As a result, robots will only have a pseudo or functional morality. I will consider definitions of artificial moral agency in both the strong and weak machine ethics perspectives.

---

[4] Can be found in the Nicomachean Ethics Book VI.

Given this context, I can now discuss my candidate definition for artificial moral agency. To do this; it is essential to understand that current approaches to machine ethics are primarily computational, i.e. they are dealing with *computational morality*. Outside of significant advances in new approaches to designing artificial agents, it seems unlikely that this will change soon. Even those that recognise that some notion of consciousness will be required for general intelligence (Franklin 2003), and indeed full moral agency (Wallach et al. 2011), are only working towards functional approximations of it—mostly using a combination of cognitive architectures and computational implementations (Franklin et al. 2014; Lucentini and Gudwin 2015). The nature of machine ethics implementations, it would seem, will remain almost certainly computational, at least for the foreseeable future.

The definition of moral agency given by Parthemore and Whitby (2014, p. 1)[5] serves as a good reference. However, the previous discussion showed that different people mean different things when they use the term 'moral agent'. What is important for researchers and designers in machine ethics is to state clearly in which sense we mean the term 'moral agent' and also clearly specify what exactly our definition for it is. To illustrate, I define artificial moral agency (in the weak sense) by modifying Parthemore and Whitby's definition as follows:

> An artificial moral agent is a computationally-based agent whom one appropriately holds responsible for its actions and consequences, and artificial moral agency is the distinct type of agency that agent possesses.

I refer to moral agency in the weak sense, meaning that I believe not all moral decisions can be made rationally—super-rational capacities are required for others. A strong machine ethics view of artificial moral agency can also be defined and clarified by following a similar process. The definition of artificial moral agency above is somewhat ontological in that it emphasises the nature of the agent. However, in theory, a definition based on the agent's moral capability could also be derived. Thankfully, Moor (2006) has already developed a taxonomy that helps characterise the level of ethical capability in artificial agents.

Moor describes four different kinds of AMA's, each according to capability. These four kinds are (in order of increasing ethical capability): *ethical impact agents*; *implicit ethical agents*; *explicit ethical agents*; and *full ethical agents*. Though a full examination of Moor's taxonomy is outside of the scope of this article, I submit that my sample definition

is quite consistent with what Moor calls a *explicit ethical agent*[6]. For the remainder of this article, I will use the sample definition of artificial moral agency stated above (in the weak sense), complemented by the use of the term *explicit ethical agent*, to be what I mean when referring to an AMA.

Artificial moral agency within a framework of computational rationality

I now need to show how the concept of artificial moral agency is compatible with a framework of computational rationality. Firstly, I will argue that the capacities necessary for moral agency lend themselves naturally to being computable. Secondly, I will also argue that many of the problems that were envisaged could be solved by computational rationality, are also present in computational morality, and that these same problems can also be solved through a framework computational rationality in the tradition of Gershman et al. (2015), Horvitz (1987), Russell and Subramanian (1995), and others. Let me begin by examining the claim that the capacities required for moral agency can be computed.

So far, I have avoided stating which capacities are required for moral agency. In the literature, these capacities can include emotions, empathy, free will, rationality, cognition (including mental and intentional states), concepts, awareness, amongst others (Wallach et al. 2011; Parthemore and Whitby 2013, 2014; Himma 2009; Torrance 2008). One way to get around this issue is to focus on what these various capacities give you as a result. In other words, instead of arguing about which capacities (and combinations thereof) will result in some facet of moral agency, focus on the outcome that is expected to be achieved. This is precisely what philosophers such as Sullins (2006) and Floridi and Sanders (2004) do by focusing on the top-level requirements for artificial moral agency and abstracting away the detail regarding the exact capacities required. I choose to focus on the requirements expressed by Floridi and Sanders because they conceptualise artificial moral agency within a weak machine ethics framework, as opposed to Sullins, who conceptualises it within a strong machine ethics framework.

Flordi and Sanders define the requirements for artificial moral agency as *interactivity* (being aware and responsive to environmental stimuli), *adaptability* (the ability to change internal states according to environmental stimuli) and *autonomy* (the ability to change internal states according to the agent's own transition rules independently of

---

[5] "A moral agent is an agent whom one appropriately holds responsible for its actions and their consequences, and moral agency is the distinct type of agency that agent possesses".

[6] According to Moor (2006), a *explicit ethical agent* is one that can hold an explicit ethical representation of a given situation, and use that to respond in a manner that is ethical.

environmental stimuli) (Floridi and Sanders 2004). Focusing on the top-level requirements for moral agency and abstracting away details around required capacities is essentially a focus on 'mindless' morality—a form of morality that distinctly suits a *computational* framing of moral agency. It does not care how autonomy or intentionality, for instance, are achieved—it only cares that they are achieved. This is precisely what Floridi and Sanders (2004) are alluding to when they talk about moral agency at different levels of abstraction (LoA). At a low enough LoA, a human being would also not be considered a moral agent since we would be dealing with their biological make-up, the neurobiological processes in their brains and other cognitive processes which at that level would seem indistinguishable from a machine.

Similarly, artificial agents observed at a low enough LoA are simply electronic components and code, and at that level, we cannot decide on moral agency. However, at a high enough LoA, these low-level processes and components are abstracted such that we only see the outcomes of their decisions. We wouldn't ordinarily know how exactly the AMA functions, only that it seems to have some goals and intentions, it can function autonomously, and learn new things over time. At that LoA, we would be forced to admit that the robot acts in a manner that is consistent with our expectations of moral agents (Coeckelbergh 2014).

Flrodi and Sanders' approach to defining the requirements for artificial moral agency is not without its critiques, the strongest of which likely comes from Himma (2009). He argues that, under Floridi and Sanders' formulation, rattlesnakes, for example, could be wrongly considered to be artificial moral agents. If, as Himma's example goes, the rattlesnake acts as a response to hunger and kills something, then it would have acted autonomously, certainly interactively and apparently with some ability to learn. The crux of Himma's argument seems to be that only praise or blameworthy agents could be moral agents. There are two issues with Himma's argument, especially as it pertains to artificial moral agency.

Firstly, Himma's argument presupposes that discourse around moral agency is equivalent to responsibility analysis and that no room exists for prescriptive discourse in the identification of moral agents (Floridi and Sanders 2004). Secondly, and to use his example, the rattlesnake would not qualify as a moral agent, according to Floridi and Sanders' requirements, because it cannot learn moral values. It is only responding to instinct.

An artificial agent, on the other hand, can be programmed to simulate the capacity to learn (morally), and thus could qualify as an AMA. How good an AMA it will be (i.e. responsibility analysis) is a different matter altogether, and will require us to build models of computational morality and to evaluate them. To be clear, without consciousness or intentional/unconscious mental states, the AMA could not be a full moral agent, but that is why we put the qualifier 'artificial' in front of 'moral agent'. In theory, it's moral performance will lie somewhere between a rattlesnake and a full moral agent such as a human being (Moor 2006).

I have argued that the capacities required for moral agency, as expressed by Floridi and Sanders (2004), lend themselves to being computable. However that is not the only reason that artificial moral agency is compatible with a framework of computational rationality. Computational rationality exists as a framework primarily because artificial agents are not perfectly rational. They face many internal and external constraints such as limited computational resources, limited information about the problem at hand, limited time (and space) within which to make a decision, the tractability of the problem itself, and so on.

As it turns out, AMA's are faced with much of the same constraints and limitations as computationally rational agents. AMA's have to make moral decisions despite the limitation of computational resources, information, time, and the tractability of the moral decision itself. I posit that the problem of computational morality is simply a special case, albeit a complex one, of computational rationality, and that many of the approaches to solving computational rationality in the general case, can be used to enhance the prospects for computational morality further.

For example, the emergence of hybrid approaches, i.e. model-based (top-down) and model-free (bottom-up), as a superior choice for certain complex tasks for computational rationality (Gershman et al. 2015), and the fact that prominent researchers in machine ethics believe that a combination of top-down and bottom-up approaches will likely be required to solve certain kinds of complex moral decisions (Allen et al. 2005), lends further credence to the idea that the two domains are more related than different[7]. Just as Russell and Subramanian (1995) popularised the concept of a bounded-optimal agent, perhaps it is time to start talking about bounded-optimal artificial moral agents, i.e. AMA's that arrive at moral decisions based on the information they can acquire from the environment, given the limitations of their software architectures and programming. Next, I will briefly discuss a basic conceptual model for a computationally rational AMA.

---

[7] The use of the terms top-down and bottom-up in both the cited philosophical and scientific disciplines is conceptually the same. In both cases, top-down means starting from a pre-defined ethical framework or a computational model and bottom-up means learning an ethical representation or a computational model from the available data.

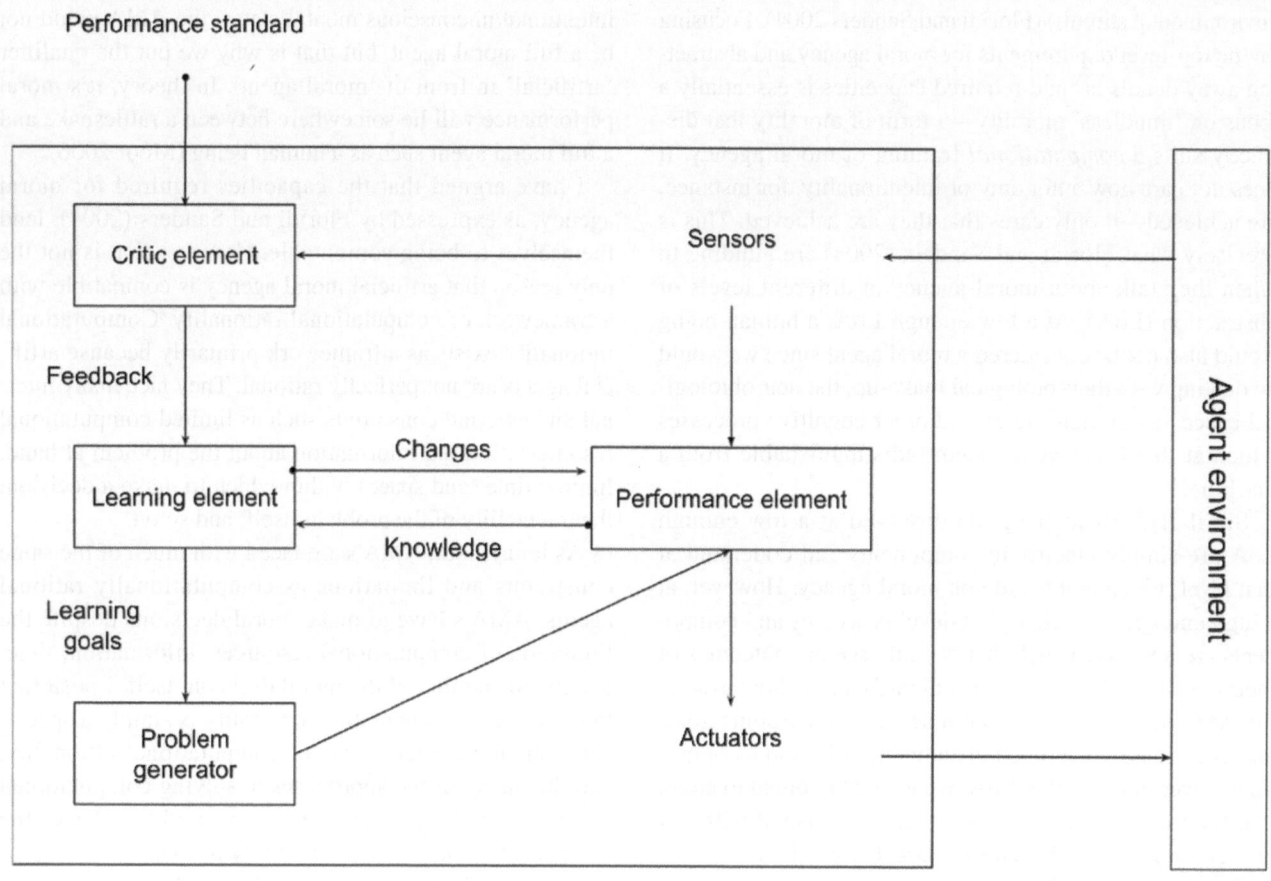

Fig. 1 A generic representaton of a general learning agent (Russell and Norvig 2009)

A model for an optimally-bounded, computationally rational AMA

The proposed model for a computationally rational AMA is based on the idea of an optimally-bounded, computationally rational agent that has been discussed thus far. I openly base the model on Russell and Norvig (2009, p. 55) (see Fig. 1), whose conception of a general learning agent is simple, and yet comprehensive. The ideas in computationally rationality can be integrated into the model of any general artificial and intelligent agent, so long as its key tenants, such as bounded-optimality, the separation of meta-reasoning from specific algorithms for reasoning, and the use of formal and heuristic methods, are preserved. Figure 1 depicts the structure of a general learning agent which can perform certain actions in an environment, through its sensors and actuators, according to a set performance standard (perhaps set by a human being). The general learning agent also can improve its decision making and performance capability over time, and generate new problems (goals) that can help it to improve performance further and learn new ways to reason.

I present Fig. 2 as a proposed high-level conceptual model for a computationally rational AMA. The agent gathers

bounded information from the environment, and processes it in the *ethical performance element*, which is responsible for ethical as well as general reasoning. The decisions and actions from this element are then transferred back to the environment (via the relevant actuators and communication mechanisms). The *learning element* and *problem generator* are left as-is from Russell and Norvig's conception. They are responsible for updating the performance element with new ways to reason and generate new ideas for future performance, respectively. The *critic element* is also similar, except instead of only allowing for external input to modify the performance of the agent (e.g. human input), it also allows the agent to provide a human-understandable rationale for its performance.

Figure 3 zooms in on the ethical performance element, where the *ethical meta-reasoner* is responsible for deciding on the best ethical framework (or combinations thereof) and one or more programs to execute to arrive at an optimally-bounded ethical decision. The ethical performance element thus separates high-level meta reasoning activities from the execution. However, it still exposes the ethical meta-reasoner to the information from the environment to allow it to make the optimal choice of execution strategy. At a high-level, the

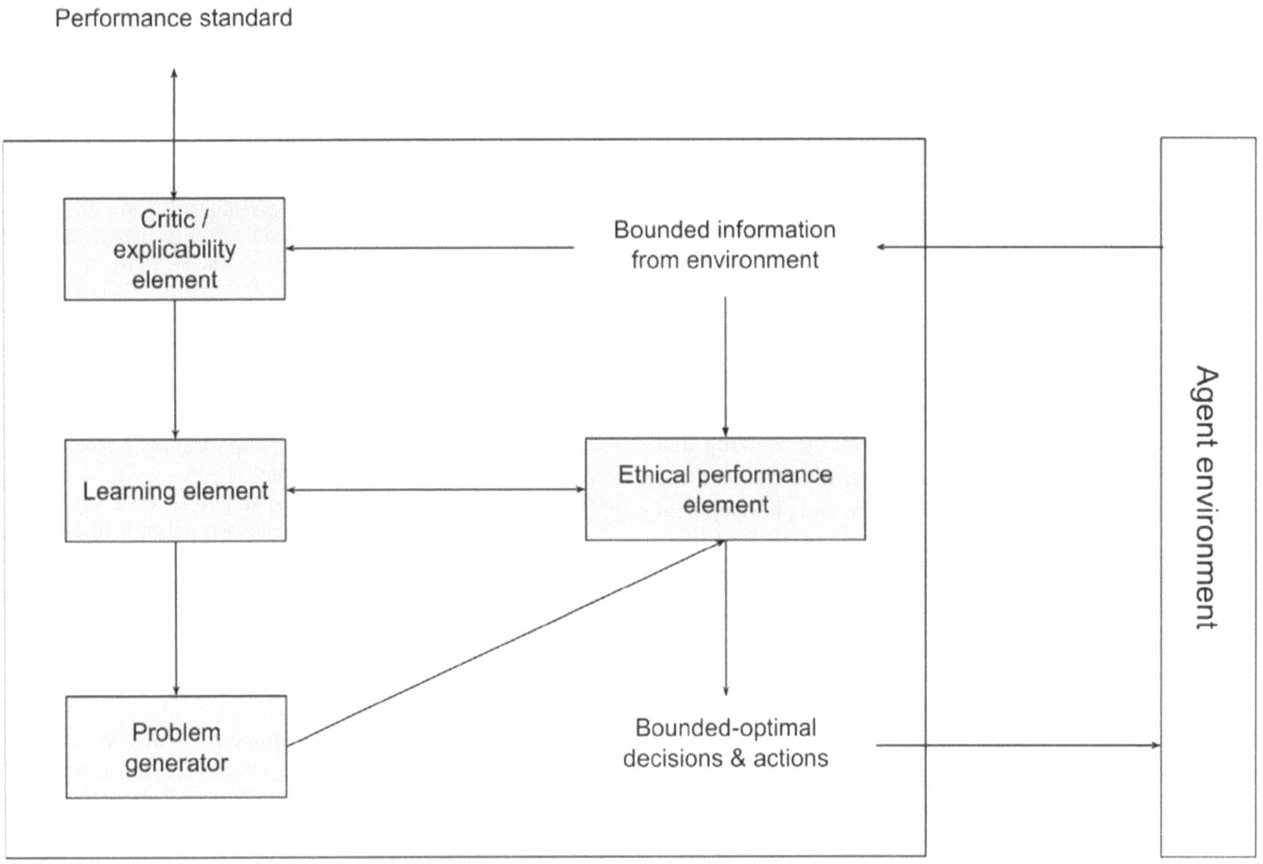

Performance standard

Critic / explicability element

Bounded information from environment

Agent environment

Learning element

Ethical performance element

Problem generator

Bounded-optimal decisions & actions

Fig. 2 A conceptual model for an optimally-bounded, computational rational AMA

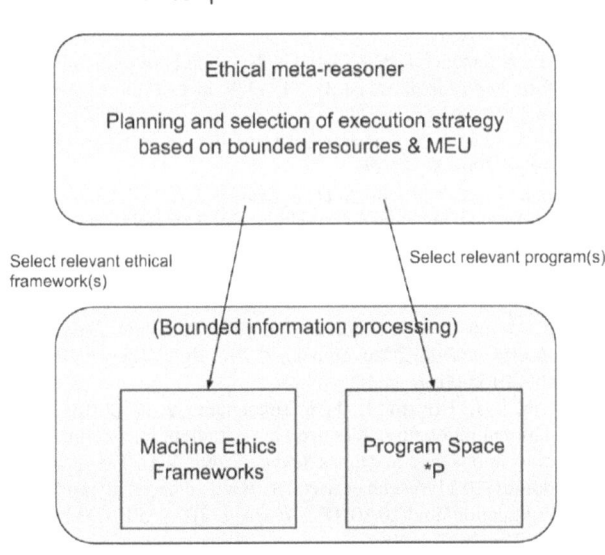

Ethical performance element

Ethical meta-reasoner

Planning and selection of execution strategy based on bounded resources & MEU

Select relevant ethical framework(s)

Select relevant program(s)

(Bounded information processing)

Machine Ethics Frameworks

Program Space *P

Fig. 3 A detailed view of the ethical performance element

proposed AMA would meet the requirements of interactivity (it can receive information from the environment and

act on it), adaptability (it can change its performance state through the ethical performance and learning elements), and autonomy (it can behave in a somewhat autonomous manner through the problem generator, which generates new ideas about how to execute performance in the future). Additionally, it can receive a new performance standard and explain its current performance to a human being.

With regards to potential limitations of the model, I have argued in earlier sections that the AMA is conceptualised to have weak machine ethics (Sect. 3). As such, we can expect that the AMA could only be capable of making some, but not all, moral decisions. At this stage, it would be difficult to determine which moral decisions it would be able to make, and such a determination lies outside the scope of this article. However, I can speculate that moral decision making in situations where (bounded) information is readily available and accessible to the AMA should be theoretically possible. Such contexts could include highly domain-specific environments, such as in self-driving cars, healthcare robots, loan approval bots, home-assistants, and the like.

The model depends heavily on the availability of bounded information. Thus I expect that moral decisions requiring little to no external information (i.e. abstract decision-making) would be difficult to compute, at least initially until the AMA learns a sufficient representation of moral values. Furthermore, there is the general issue (not necessarily a limitation, but an unknown) of how the model would internally represent its learned moral values, and how this would map to actions that affect real agents in the real world.

Conclusion

The purpose of this article was to advance an argument and model for artificial moral agency based on a framework of computational rationality. This was done by showing that computational rationality can be an integrative element that can effectively combine both the scientific and philosophical elements of artificial moral agency consistently and logically. In particular, I argued that the capacities required for artificial moral agency, as well as the aspects of functional consciousness that underpin them, are computable. I further argued that computational morality is a special, if not complex, case of computational rationality, hence many techniques originally developed for general rationality can be adapted for computational morality. I then briefly proposed a conceptual model for a bounded-optimal, computationally rational AMA.

Some philosophers and scientists might reject the idea of a bounded-optimal artificial moral agent. After all, the stakes can be quite high when it comes to moral decision making, as the wrong decision could have significant moral and societal implications. However, we need to start somewhere, and I suggest that starting from a weak machine ethics perspective is helpful to allow us to begin to test its limits and the sorts of domains and contexts where it can be applied. The model proposed is an invitation for dialogue and feedback, and the hope is that many philosopher-developers pairs can be formed to solve the problem of constraining weak AI systems and making them more respecting of human moral values.

I have specifically chosen to omit mentions of the ethical frameworks that the AMA should follow, as the main purpose of this article was to locate artificial moral agency within a framework of computational rationality. Future research needs to focus on the kinds of ethical frameworks an optimally-bounded, computationally rational AMA ought to follow. Further research into appropriate software architectures for the AMA, and the type of programs that can form part of the ethical performance element's program space, is also required.

References

Abney, K. (2012). Robotics, ethical theory, and metaethics: A guide for the perplexed, chap 3. In P. Lin, K. Abney, & G. Bekey (Eds.), *Robot Ethics, the ethical and social implications of robotics*. Cambridge: The MIT Press.

Allen, C., & Wallach, W. (2012). Moral Machines: contradiction in terms, or abdication of human responsibility? Chap 4. In P. Lin, K. Abney, & G. A. Bekey (Eds.), *Robot Ethics, the ethical and social implications of robotics*. Cambridge: The MIT Press.

Allen, C., Smit, I., & Wallach, W. (2005). Artificial morality: Top-down, bottom-up, and hybrid approaches. *Ethics and Information Technology*, 7(3), 149–155. https://doi.org/10.1007/s1067 6-006-0004-4.

Anderson, M., & Anderson, S. L. (2007). Machine ethics: Creating an ethical intelligent agent. *AI Magazine*, 28(4), 15. https://doi.org/10.1609/aimag.v28i4.2065, http://www.aaai.org/ojs/index.php/aimagazine/article/view/2065.

Arnold, T., Kasenberg, D., & Scheutz, M. (2017). Value alignment or misalignment what will keep systems accountable?. In *Workshops at the Thirty-First AAAI Conference on Artificial Intelligence*.

Churchland, P. S. (2014). The neurobiological platform for moral values. *Behaviour*, 151(2–3), 283–296. https://doi.org/10.1163/1568539X-00003144.

Coeckelbergh, M. (2014). The moral standing of machines: Towards a relational and non-cartesian moral hermeneutics. *Philosophy and Technology*, 27(1), 61–77.

Conitzer, V., Sinnott-Armstrong, W., Borg, J. S., Deng, Y., & Kramer, M. (2017). Moral decision making frameworks for artificial intelligence. In *Thirty-First AAAI Conference on Artificial Intelligence*, https://pdfs.semanticscholar.org/a3bb/ffdcc1c7c4cae66d6af373651389d94b7090.pdf.

Daily, M., Medasani, S., Behringer, R., & Trivedi, M. (2017). Self-driving cars. *Computer*, 50(12), 18–23. https://doi.org/10.1109/MC.2017.4451204

Dameski, A. (2018). A comprehensive ethical framework for AI entities: Foundations. In M. Iklé, A. Franz, R. Rzepka, B. Goertzel, (Eds.), *International Conference on Artificial General Intelligence*, pp. 42–51. Berlin: Springer. https://doi.org/10.1007/978-3-319-97676-1.

Floridi, L., & Sanders, J. W. (2004). On the morality of artificial agents. *Minds and Machines*, 14(3), 349–379. https://doi.org/10.2139/ssrn.1124296.

Franklin, S. (2003). A conscious artifact? *Journal of Consciousness Studies*, 10(4–5), 47–66.

Franklin, S., Madl, T., Mello, S. D., & Snaider, J. (2014). LIDA: A systems-level architecture for cognition, emotion, and learning. *IEEE Transactions on Autonomous Mental Development*, 6(1), 19–41.

Genewein, T., Leibfried, F., Grau-Moya, J., & Braun, D. A. (2015). Bounded rationality, abstraction, and hierarchical decision-making: An information-theoretic optimality principle. *Frontiers in Robotics and AI*, 2(November), 1–24. https://doi.org/10.3389/frobt.2015.00027.

Gershman, S. J., Horvitz, E. J., & Tenenbaum, J. B. (2015). Computational rationality: A converging paradigm for intelligence in brains, minds, and machines. *Science*, 349(6245), 273–278. https://doi.org/10.1126/science.aac6076, www.sciencemag.orgpapers2://publication/uuid/20A0106C-9CBA-472D-AAFB-69231964766F, arXiv:1011.1669v3.

Himma, K. E. (2009). Artificial agency, consciousness, and the criteria for moral agency: What properties must an artificial agent have to be a moral agent? *Ethics and Information Technology*, 11(1), 19–29. https://doi.org/10.1007/s10676-008-9167-5.

Horvitz, E. J. (1987). Reasoning about beliefs and actions under computational resource constraints. In *Proceedings of the Third*

Workshop on Uncertainty in Artificial Intelligence, AAAI and Association for Uncertainty in Artificial Intelligence, pp. 429–444. http://erichorvitz.com/u87.htm.

Horvitz, E. J. (1988). Reasoning under varying and uncertain resource constraints. In *AAAI*, pp. 111–116.

Horvitz, E. J. (1989). Rational metareasoning and compilation for optimizing decisions under bounded resources. In *Proceedings of Computational Intelligence '89*, Association of Computing Machinery, Milan, Italy, http://erichorvitz.com/rationality_89.htm.

Horvitz, E. J., Cooper, G. F., & Heckerman, D. E. (1989). Reflection and action under scarce resources: Theoretical principles and empirical study. *IJCAI*, 2, 1121–1127.

Jiang, F., Jiang, Y., Zhi, H., Dong, Y., Li, H., Ma, S., et al. (2017). Artificial intelligence in healthcare: past, present and future. *BMJ*,. https://doi.org/10.1136/svn-2017-000101.

Johnson, D. G. (2006). Computer systems: Moral entities but not moral agents. *Machine Ethics*, 9780521112, 168–183. https://doi.org/10.1017/CBO9780511978036.012.

Leviathan, Y., & Matias, Y. (2017). Google AI Blog: Google Duplex: An AI system for accomplishing real-world tasks over the phone. https://ai.googleblog.com/2018/05/duplex-ai-system-for-natural-conversation.html.

Lewis, R. L., Howes, A., & Singh, S. (2014). Computational rationality: Linking mechanism and behavior through bounded utility maximization. *Topics in Cognitive Science*,. https://doi.org/10.1111/tops.12086.

Liao, S. M. (2010). The basis of human moral status. *Journal of Moral Philosophy*, 7(2), 1–31. https://doi.org/10.1163/174552409X12567397529106.

Lucentini, D. F., & Gudwin, R. R. (2015). A comparison among cognitive architectures: A theoretical analysis. *Procedia Procedia Computer Science*, 71, 56–61. https://doi.org/10.1016/j.procs.2015.12.198.

Marwala, T. (2013). Semi-bounded rationality—A model for decision making. arXiv preprint arXiv:13056037 pp. 153–164, arXiv:1305.6037.

McCarthy, J., Minsky, M. L., Rochester, N., & Shannon, C. E. (2006). A proposal for the Dartmouth summer research project on artificial intelligence. *AI Magazine*, 4, 12–14. https://doi.org/10.1609/aimag.v27i4.1904. arXiv:9809069v1.

Miller, F. D. (1984). Aristotle on rationality in action. *The Review of Metaphysics*, 37(3), 499–520, https://www.jstor.org/stable/20128047.

Moor, J. H. (2006). The nature, importance, and difficulty of machine ethics. *IEEE Intelligent Systems*, 21(4), 18–21. https://doi.org/10.1109/MIS.2006.80.

Parthemore, J., & Whitby, B. (2013). What makes any agent a moral agent? Reflections on machine consciousness and moral Agency. *International Journal of Machine Consciousness*, 5(2), 105–129. https://pdfs.semanticscholar.org/3ff2/49fe3c8b3a2c94ae762b76b2dd0203f1f789.pdf.

Parthemore, J., & Whitby, B. (2014). Moral agency, moral responsibility, and artifacts: What existing artifacts fail to achieve (and why), and why they, nevertheless, can (and do!) make moral claims upon us. *International Journal of Machine Consciousness*, 6(2), 141–161. https://doi.org/10.1142/S1793843014400162.

Rottschaefer, W. A. (2000). Naturalizing ethics: The biology and psychology of moral agency. *Zygon*, 35(5–6), 253–286. https://doi.org/10.1111/0591-2385.00276.

Russell, S. J., & Norvig, P. (2009). *Artifical intelligence: A modern approach*, third edit edn. Prentice Hall, https://doi.org/10.1017/S0269888900007724, arXiv:1707.02286, arXiv:1011.1669v3.

Russell, S. J., & Subramanian, D. (1995). Provably bounded-optimal agents. *Journal of Artiicial Intelligence Research*, 2, 575–609.

Sapaty, P. S. (2015). Military robotics: Latest trends and spatial grasp solutions. *IJARAI International Journal of Advanced Research in Artificial Intelligence*, 4(4), 9–18.

Scheutz, M., & Malle, B. F. (2017). Moral robots. In L. S. M. Johnson & K. S. Rommelfanger (Eds.), *The Routledge handbook of neuroethics*. Abington: Routledge. https://doi.org/10.4324/9781315708652.ch24.

Schlosser, M. (2015). Agency. In E. N. Zalta (Ed.), *The Stanford encyclopedia of philosophy, fall 2015 edition*. Stanford: Metaphysics Research Lab, Stanford University.

Selten, R. (1990). Bounded rationality. *Journal of Institutional and Theoretical Economics (JITE)*, 146(4), 649–658.

Silver, D., Schrittwieser, J., Simonyan, K., Antonoglou, I., Huang, A., Guez, A., et al. (2017). Mastering the game of Go without human knowledge. *Nature*, 550(7676), 354. https://doi.org/10.1038/nature24270.

Simon, H. A. (1955). A behavioral model of rational choice. *The Quarterly Journal of Economics*, 69(1), 99–118.

Simon, H. A. (1972). Theories of bounded rationality. *Decision and Organization*, 1(1), 161–176.

Sullins, J. P. (2006). When is a robot a moral agent? IRIE: International Review of Information Ethics. http://sonoma-dspace.calstate.edu/handle/10211.1/427.

Torrance, S. (2008). Ethics and consciousness in artificial agents. *AI and Society*, 22(4), 495–521. https://doi.org/10.1007/s00146-007-0091-8.

Torrance, S. (2013). Artificial agents and the expanding ethical circle. *AI and Society*, 28(4), 399–414. https://doi.org/10.1007/s00146-012-0422-2.

Turing, A. (1950). Computing machinery and intelligence. *Mind*, 59(236), 433–460.

Wallach, W., Allen, C., & Franklin, S. (2011). Consciousness and ethics: Artificially conscious moral agents. *International Journal of Machine Consciousness*, 03(01), 177–192. https://doi.org/10.1142/S1793843011000674.

Wu, Y. H., & Lin, S. D. (2018). A low-cost ethics shaping approach for designing reinforcement learning agents. *The Thirty-Second AAAI Conference on Artificial Intelligence (AAAI-18)*. arXiv:1712.04172.

Yu, H., Shen, Z., Miao, C., Leung, C., Lesser, V. R., & Yang, Q. (2018). Building ethics into artificial intelligence. *Proceedings of the Twenty-Seventh International Joint Conference on Artificial Intelligence (IJCAI)*, pp. 5527–5533. http://moralmachine.mit.edu/.

Zilberstein, S. (2013). Metareasoning and bounded rationality. In M. T. Cox & A. Raja (Eds.), *Metareasoning: Thinking about thinking* (pp. 27–40). Cambridge: MIT Press. https://doi.org/10.7551/mitpress/9780262014809.003.0003.

Ethics and Information Technology (2021) 23:147–155
https://doi.org/10.1007/s10676-020-09540-4

ORIGINAL PAPER

The artificial view: toward a non-anthropocentric account of moral patiency

Fabio Tollon[1]

Published online: 1 June 2020
© Springer Nature B.V. 2020

Abstract
In this paper I provide an exposition and critique of the Organic View of Ethical Status, as outlined by Torrance (2008). A key presupposition of this view is that only moral patients can be moral agents. It is claimed that because artificial agents lack sentience, they cannot be proper subjects of moral concern (i.e. moral patients). This account of moral standing in principle excludes machines from participating in our moral universe. I will argue that the Organic View operationalises anthropocentric intuitions regarding sentience ascription, and by extension how we identify moral patients. The main difference between the argument I provide here and traditional arguments surrounding moral attributability is that I do not necessarily defend the view that internal states ground our ascriptions of moral patiency. This is in contrast to views such as those defended by Singer (1975, 2011) and Torrance (2008), where concepts such as sentience play starring roles. I will raise both conceptual and epistemic issues with regards to this sense of sentience. While this does not preclude the usage of sentience outright, it suggests that we should be more careful in our usage of internal mental states to ground our moral ascriptions. Following from this I suggest other avenues for further exploration into machine moral patiency which may not have the same shortcomings as the Organic View.

Keywords Machine moral patiency · Sentience · Anthropocentrism · Intentional stance · Organic view of ethical status

Introduction

When evaluating moral situations, we tend to think in terms of giving moral stakeholders their due: giving them what they *deserve* based either on how they have behaved or whether they have been harmed. There arises, firstly, the question of whether an entity is misbehaving *intentionally*, in the common-sense usage of the term ("on purpose"), and whether it could in some sense be *responsible* for its behaviour, and hence possibly morally responsible. This is a question of moral *agency*. Conversely, a second question may arise of whether, if we were to harm the entity, we would be doing it a *moral harm*. In other words, do we owe it certain moral *obligations*? This is a question of moral *patiency*. These two questions can be viewed as fundamental to all moral philosophy: *who* or *what* is deserving of moral concern, and *who* or *what* can be said to be (morally)

responsible for their actions (Gunkel 2012, p. 1). Moral patients are the class of entities that can in principle qualify as *receivers* of moral action, whereas moral agents are the class of entities that can in principle qualify as *sources* of moral action (Floridi and Sanders 2004, pp. 349–350).

On the one hand, trends in contemporary macro-ethics have been geared toward expanding the boundaries of moral consideration by focusing on the nature of who or what should count as a moral patient. This ascription of moral patiency is independent of whether the entity in question is a moral agent or not (Floridi and Sanders 2004).[1] However, while not all moral patients are moral agents, it is standardly supposed that all moral agents are moral patients (see Floridi and Sanders 2004; Torrance 2008). On the other hand, the

✉ Fabio Tollon
  fabiotollon@gmail.com

1 Philosophy Department, Stellenbosch University,
   Stellenbosch, Western Cape, South Africa

[1] A patient-orientated approach to ethics is not concerned with the perpetrator of a specific action, but rather attempts to zero in on the *victim* or receiver of the action (Floridi, 1999). This type of approach to ethics is considered non-standard and has been incredibly influential in both the "animal liberation" movement and "deep ecology" approaches to environmentalism (see Leopold, 1948; Naess, 1973; Singer, 1975, 2011). Both place an emphasis on the *victims* of moral harms; in the case of animal liberation, the harm we do to animals, and in the case of deep ecology the harm we do to the environment.

Chapter 6 was originally published as Tollon, F. Ethics and Information Technology (2021) 23: 147–155. https://doi.org/10.1007/s10676-020-09540-4.

emergence of artificially intelligent systems, properly conceptualised as artificial agents[2] (AAs), may complicate many presuppositions of what counts as a moral action. These systems may come to undermine the standard assumption above by performing actions which, while independent of human control, might still be subject to moral assessment (see Sparrow 2007; Johansson 2010; cf. Johnson and Noorman 2014; Johnson 2015). While the latter question is deserving of (and has received) considerable philosophical attention, my focus in this paper will not be concerned with moral agency directly. Instead, I will assume the validity of the conceptual relationship between agents and patients which claims that all moral agents are moral patients. It is with the aforementioned in mind that any investigation into moral agency must first address the question of moral patiency. This conceptual point stresses the importance of the discussion in this paper, as the implications of this approach for machines are clear: if machines cannot be considered moral patients, then they cannot be moral agents either. The stakes in this debate are quite high. If we were to conclude that no computationally-based systems can ever be fitting subjects of moral concern, then our treatment of them need not follow any moral contours. Our treating them and their needs as morally subordinate to our own would not be problematic, as we would owe them no moral obligations. However, if it turns out that we were wrong to treat these systems as "mere machines", then we would find ourselves guilty of harming an entirely new class of moral patient, and unjustifiably excluding them from our moral universe. It is therefore important that we take the "machine question"[3] seriously (Gunkel 2012, p. 5).

The organic view of ethical status

In order to address the question of machine moral patiency I will provide an exposition and critique of the Organic View of Ethical Status (hereafter simply the "Organic View"), as it is articulated by Steve Torrance (2008). The Organic View makes an important contribution to the philosophical debate on moral status. Torrance's exposition of the Organic View brings together many characteristics that make a consistent

appearance in the literature on machine moral agency and patiency. These are questions of sentience, intentionality, and the conceptual relationship between moral agents and moral patients (see Floridi and Sanders 2004; Johnson and Miller 2008; Himma 2009; Sullins 2011; Johnson and Noorman 2014).

The Organic View raises pertinent ethical questions, specifically, whether the expansion of our "mental" universe to include machines also necessitates an expansion of our moral universe to include them (Torrance 2013, p. 399). In order to make his case, Torrance centres his discussion around two factors which feature prominently in the Organic View: firstly, he claims that *sentience*[4] (or phenomenal consciousness) is a key factor in the type of rationality moral entities exhibit, and, secondly, that *biological constitution* is of fundamental moral significance (2008, p. 505). This paper focuses on the first of these claims, and while Torrance does not explicitly endorse the Organic View, he does seem to harbour a favourable disposition toward it. While he is willing to concede that it may well be wrong (or at the very least in need of further qualification) (Torrance 2008, p. 505), I will endeavour to show, in line with the work of Mark Coeckelbergh, that the Organic View succumbs to issues of *justification* in terms of moral consideration (2010a,2014; b). The content of this paper is therefore broadly in line with Coeckelbergh's project: it takes seriously the expansion of our moral universe in a way that does not rely only on ontological features of the entity in question (2010, p. 212). I will show how the Organic View gives us a philosophically interesting way in which to view the moral status of artificial systems, but that it nonetheless still falls victim to the issues raised by Coeckelbergh (2010a, b 2014). In order to make my argument I first put forward the case made by Torrance (2008, p. 503) that AAs do not have "empathic rationality", with the implication that machines, unless they can be designated as "sentient", cannot be proper subjects of moral concern. From this, I then show how the sense of sentience Torrance operationalises in his account is flawed due to both conceptual and epistemic shortcomings.

Empathic rationality

In this section I deal with a specific (but essential) claim of the Organic View: "Only beings which are capable of sentient feeling or phenomenal awareness could be genuine subjects of either moral concern or moral appraisal" (ibid., p. 503). The reason for focusing on this aspect of the Organic View is that, if found wanting, it would undermine the entire

---

[2] An artificial agent is artificial in the sense that it has been manufactured by intentional agents (us) out of pre-existing materials, which are external to the manufacturers themselves (Himma, 2009). It is an agent in the sense that it is capable of performing actions (Floridi and Sanders, 2004: 349). An easy example of such an artificial agent would be a cellphone, as it is manufactured by humans and can perform actions, such as basic arithmetic functions or responding to queries via online searches.

[3] Gunkel (2012: 5) considers the "machine question" to be the flip side of the "animal question": both concern the moral standing of non-human entities.

[4] Sentience can be understood as the capacity for an entity to have phenomenal/subjective/qualitative states of experience (Bostrom and Yudkowsky, 2011: 7).

argument. The criterion of sentience is what grounds Torrance's conception of moral patiency, and so if it can be found wanting it would be a serious threat to the validity of the argument. This will become clear as my critique develops.

Torrance begins his argument by asking us to imagine an AA that has a certain minimum level of rationality and has the cognitive ability to recognise that certain beings have sentient states, and thus moral interests (Torrance 2008, p. 510). Moreover, the AA can reason about the effects that different courses of action may have on these sentient creatures. Yet, this type of agent does not have the capacity to *feel* moral concern (ibid.). Such agents, due to their ability to *cognitively* apprehend and interpret the behavioural cues of other entities, and to infer from these that the entity in question could be undergoing a moral harm, etc., *may* be thought of as being fitting subjects of moral appraisal (ibid.). Due to their ability to cognitively apprehend and reason about moral situations, these entities could use this ability to guide their actions – these then being subject to moral evaluation. In other words, by fulfilling certain rationality criteria (which other non-human entities do not), one might think it reasonable to extend the ascription of moral agency to these entities; even if they are not sentient in the same way that human beings are (ibid.).

Nevertheless, the problem with this view, according to Torrance (ibid.), lies in assuming that the type of rationality required for moral agency is simply cognitive or intellectual, as this would provide us with an anaemic account of moral standing. Torrance suggests that the kind of rationality required for an entity to legitimately be given the status of moral agent may turn out to be different from the kind that could be achieved by an AI system. He argues that the type of rationality traditionally associated with our own full moral status (as humans) is closely associated with our sentient nature (in other words, our capacity for *affect*) (ibid.). Thus the claim is that being a moral agent requires (human) sentience (or affect) (ibid.). The argument goes as follows: our kind of rationality involves the capacity for a kind of affective or empathetic identification with the experiential states of others, where such identification is integrally available to the agent as an essential component in its moral decision-making procedures (ibid.). Torrance (ibid.: 516) calls this kind of rationality *empathic rationality* and contrasts it with the purely *cognitive* or *intellectual rationality*, which might be attributable to intelligent, computationally-based AAs. While we expect information-processing systems to make decisions in a purely mechanistic way, Torrance claims that we have different standards when it comes to our moral decision-making procedures, as we expect human beings to factor the potential experiential consequences of their actions into their moral reasoning (ibid., p. 511). Significantly, he claims that entities which are only capable

of intellectual rationality would not have a "real" or "true" means of evaluating the experiential states of others. Such an entity could simply not understand how its actions might affect others.

Thus, Torrance's argument is that moral decision making requires the capacity for "engaged empathic rational reflection" (ibid., p. 511), which requires the ability to identify with the experiential states of others. Any rational agent that is not also sentient (in a manner equivalent to the type of sentience achievable by biological organisms) would not have this empathic ability, since a precondition for a "true" understanding of experiential states is that one is able to have these states oneself. Since only entities capable of being "ethical consumers" can have this type of empathic rationality, other types of agents are precluded from being subject to moral evaluation, as without the ability to take a "moral point of view", it would be a mistake to then evaluate actions undertaken by such agents using moral criteria (ibid., p. 499). The Organic View suggests, then, that we should conclude that entities lacking a specific type of sentience cannot be moral agents.

Problems with the organic view of ethical status

The first ambiguity that needs to be addressed is the vague way in which internal, experiential states are operationalised in Torrance's articulation of the Organic View. Only organisms capable of having some kind of "qualitative experience" of pain (or any other such experiential state) will qualify as moral patients (and by extension, according to the Organic View, as moral agents).[5] Moreover, as Torrance (2014) is a realist about mental states, he claims that there is an *objective* answer when we ask the question as to an entity's psychological state.[6] This realism about mental states works to buttress his views regarding our moral ascriptions to artificial entities: Torrance's specific form of realism claims that even if there were no functional or cognitive difference between an artificial and biological system, there

---

[5] For the sake of argument, I focus here on the experience of pain, but logically it would be possible to subject any type of internal mental state to the same type of analysis. Any theory which posits an "experience of X" claim must eventually answer to the question of *who* or *what* (i.e. what *type of mind*) is *experiencing*, or capable of experiencing, X.

[6] Torrance does not believe that functionalist accounts of mind fully capture the qualitative aspects of experience. He thus believes in the metaphysical possibility of "philosophical zombies"; humans which look and behave indistinguishably from us but lack phenomenal conscious states of experience (Torrance, 2008). This is a thorny philosophical issue in its own right, but I will not go into further detail here.

would still be a *phenomenal*[7] difference (ibid., p. 13).[8] This phenomenal difference is of fundamental moral significance for Torrance given that he claims some biological form of sentience is a prerequisite for moral patiency. In what follows, I will, firstly, bring to light conceptual ambiguities inherent to the Organic View, and secondly, discuss how the epistemic distinction between the mere "appearance" of something and the "real thing" operationalised in the Organic View is a problematic one.

Conceptual Issues

To see the ambiguity more clearly, an example put forward by Daniel Dennett (1996) offers a wonderful (albeit grisly) illustration of this. Dennett asks us to imagine that:

A man's arm has been cut off in a terrible accident, but the surgeons think they can reattach it. While it is lying there, still soft and warm, on the operating table, does it feel pain? A silly suggestion you reply; it takes a mind to feel pain, and as long as the arm is not attached to a body with a mind, whatever you do to the arm can't cause suffering in any mind. (ibid., pp. 16–17)

Our intuition is that, although it might be possible to argue that the detached arm on the table may be capable of adverse nerve stimulus (i.e. pain), without being attached to some kind of mind this pain can never constitute suffering. The *experience* of pain is equivalent to suffering, and without an *experiencer* pain in itself can be of no moral significance (Gunkel 2012, p. 115). At this point a defender of the Organic View can agree, as this seems to be the exact point that they are arguing for, as only *genuinely* sentient creatures would be deserving of moral concern. Such sentient creatures are the equivalent of an "experiencer of pain" in the example above, in that they are the "experiencers of moral violation"; however, in what follows I will argue that this is a problematic stance to adopt.

While it might be reasonable to attribute the status of moral patient to certain classes of sentient animals, as we go further down the phylogenetic tree, and as creatures differ

from us in their external appearance, we tend to be less likely to attribute the requisite kind of sentience to them. Torrance acknowledges this issue but remains neutral on whether this is a strength or a weakness of the Organic View (2008, p. 515). I believe it is a weakness of the position, as one of the main reasons for the issue arising in the first place is Torrance's emphasis on biological sentience as a key ontological property of entities deserving of moral concern. As Gunkel points out, there seems to be an "irreducible terminological slippage associated with this concept" (2012, p. 115). Moreover, "suffering easily becomes conflated with and a surrogate for consciousness and mind" (Gunkel 2012, p. 115). We are inclined to view other *hominids* as sentient, but most would not award this same ascription to other creatures which perhaps have more "basic" minds, such as molluscs. We tend to think of them as analogous to the arm on the table: capable, perhaps, of adverse nerve stimulus, but not *sentient* to the required degree, not capable of *experiencing* pain. Moreover, the Organic View itself does not give us a clear criterion for sentience (of the requisite kind), and so we have to rely on our intuitions to determine which kinds of creatures are moral patients, and these intuitions are geared towards including those entities that look like us and excluding those that look less like us.

These intuitions do not necessarily track "actual" sentience, and so the criterion of sentience does not help us, in practice, to identify moral patients. Gunkel (2012) makes a similar point when discussing the various issues surrounding our identification of suffering and pain in animals. He states that while it *seems* our intuitive ascriptions make sense, we still do not have a settled answer to the question for what distinguishes pain from suffering (Gunkel 2012, p. 115). To see this more clearly consider the example of fish, more specifically, fish cognition. Our *perception* of an animal's intelligence is often a key criterion (although not the only one) for whether we consider them to be sentient or not, and fish are rarely considered to be intelligent or phenomenally sentient in a manner akin to humans or even mammals. Moreover, fish are very rarely (if ever) accorded the same type of moral concern as are warm-blooded, non-human animals. Standard reasons given for such claims is that fish lack the requisite neural complexity in order to have the right kind of "experience". Such endothermism[9] (in the case of fish, specifically) stems from a disjunction between the public perception of fish intelligence and scientific reality (Brown 2015). There is ample scientific evidence supporting the conclusion that "fish perception and cognitive abilities often match or exceed other vertebrates" (*ibid.*). For example, fish are capable of tool use and display evidence of complex social organisation and interaction (such

---

[7] Phenomenal in the sense of having the capacity for conscious awareness. When applied to his argument for moral status, however, Torrance does not require that the entity in question be self-aware, only sentient (2008: 503).

[8] My own view is that there is in fact no difference between what can be "functionally" known about the mind and "phenomenal" aspects of mind: the phenomenal is just a special case of the functional, and in this way, there is no "hard problem" of consciousness. See Chalmers (1996) for a defense of the hard problem, and Cohen and Dennett (2011) for a substantive critique.

---

[9] That is, unfair moral discrimination based on the temperature of an entity's blood.

as signs of cooperation and reconciliation) (*ibid.*). The point here is not to outline all of the ways in which fish cognition may be measured. Rather, the key issue is that if we use our traditional metrics of intelligence when it comes to animals (such as tool use and social organisation), then we are forced to conclude that fish are on par with (and at times exceed) other "sentient" vertebrates in these criteria. The next question, then, would be whether, following from the fact that fish exhibit "intelligent" behaviour, they are also phenomenally sentient and hence capable of similar kinds of suffering? Our intuitions surrounding fish sentience and their capacity to feel and suffer seem to be biased away from accepting them as sentient "enough" to merit moral concern. It seems that we struggle to empathise with fish as.

> [w]e cannot hear them vocalise, and they lack recognisable facial expressions both of which are primary cues for human empathy. Because we are not familiar with them, we do not notice behavioural signs indicative of poor welfare. (*ibid.*)

This implies that a proper, scientific construal of fish behaviour would support the conclusion that fish have relatively complex cognitive capacities, are capable of suffering, and are therefore sentient in a manner similar to creatures that are accorded moral concern (ibid.). To bring this back to the Organic View, the issue that the example above was meant to highlight is that how we go about identifying moral patients should not be guided by concepts with have intractable conceptual slippage associated with their usage.

Applying the discussion above to the question of whether an artificial system could, in principle, be the subject of moral concern, highlights the potential for moral harm in the future. In the same way that we have biases that cause us to accord a lesser moral status to non-human entities that do not sufficiently look like us, we may be biased against machines based on their unfamiliar appearance. This is not to claim that sentience can have no purchase whatsoever when it comes to moral ascription, but rather to assert that the vague description of sentience used in the Organic View, on my reading, provides an *anthropocentrically biased* understanding of what *constitutes* sentience in the first place. Even within biological species we still struggle to accurately discriminate between creatures that are "genuinely" capable of affect or not, often relying on anthropocentric intuitions instead of argument, as noted in the example of fish cognition above.

Epistemic issues

The second complication to be unpacked is the distinction between a mere ersatz phenomenon and its "true" instantiation. This is an idea which has a considerable amount of

philosophical baggage, has been around since at least Plato, and which is a recurring theme throughout the Western philosophical canon (Gunkel 2012, p. 138). By making use of sentience as the underlying capacity which qualifies/disqualifies an entity as having a moral stake, what the Organic View is in fact claiming is that only entities with the *real* capacity for phenomenal states qualify: the mere appearance of behavioural cues that point to phenomenal states (as may be the case with anthropomorphic robots) is not enough to ground our moral ascriptions, and as such only entities that are *genuinely* sentient can be accorded a moral stake. Moreover, Torrance also claims that the type of consciousness that should ground a coherent account of moral status should track a "thick" conception of phenomenality (Torrance 2007). On this "thick" conception of phenomenality a person's consciousness is deeply embodied ("lived embodiment", to use Torrance's phrasing) and inseparable from everything about that person (ibid., p. 160). This is in contrast to "thin" conceptions of phenomenality, which Torrance takes issue with. "Thin" conceptions tend to view consciousness as something that can be detached from the entity in question. The key question then becomes how we are to go about recognizing whether entities are phenomenally conscious (sentient) in this "thick" sense.

How exactly are we to go about "proving" that an organism is sentient, *really* sentient (i.e. "phenomenally conscious")? As Dennett derisively points out, "everybody agrees that sentience requires sensitivity plus some further as yet unidentified 'factor x'" (1996, p. 66). Considering my discussion above regarding how we conceptualise sentience in non-human creatures, how are we to make an epistemically sound judgment as to what counts as, for example, "real" pain versus the mere "appearance" of pain? The fuzzy nature of the concept being employed (sentience) renders it immune to such an analysis.

To see how this might be the case, consider the classic British television game show *Would I Lie to You?* In the show, contestants are split into two teams, competing against one another in attempts at deception. In each round, one contestant from each team is randomly selected and reads aloud from a card with a note on it. The content of the note is unknown to the contestants until they read it, and the goal for the contestant who has read the card out loud is to convince their opponents that what has been read is in fact the truth. The content on the cards is of a personal nature, and so only the contestant who is reading the card will know whether it is the truth or not: the opponents have no idea and are allowed to ask probing questions, which the speaker must attempt to answer in a believable way. Once the questioning is over, the opposing team can decide to either claim that they believe the speaker to be telling the truth, or claim that they believe them to have lied. After they have submitted their decision, the speaker reveals whether the note was in fact a truth or a lie, and if the opponents guessed correctly, they receive a point.

While British television can be as dry as academic philosophy, that is not the point I wish to make. In the case of the game show there is a type of deception at play: the speaker is attempting to convince the other team of the truth or falsity of their note. Similarly, when discussing questions of true sentience versus ersatz-sentience, we are attempting to figure who or what is on either side of the divide. We interpret the available evidence and then need to come to a sound judgment about the entity in question. However, and this point is crucial, in the case of the game the deception is removed: we are shown the veracity of the matter when the speaker reveals whether they were telling the truth or not. In the case of our sentience ascriptions, we have no such epistemic security: we do not have the privileged access required in order to know whether we have made the correct judgment or not. There is no verifiable test we can perform in order to determine whether we have made the correct kind of ascription. The reason we have these issues is due to a kind of epistemic opacity—we do not have direct access to the qualitative states of others and are therefore not in a good position to judge whether an entity is "truly" sentient or not.[10] While the Organic View does not explicitly require such access to these states, it nonetheless remains an open question as to how we are to know whether the entity in question is sentient. It seems as though we would also require further evidence which would, for example, show that there is indeed a *causal connection* between being a biological organism and sentience.

Moreover, as argued by Gunkel, any inference about internal states made from various external cues requires a "leap of faith" (2012). This leap, according to Gunkel, is not properly defined nor easily defendable in each and every case. An example he uses is that of a cat that screams in pain versus a lobster which is being boiled. Our intuitions suggest to us that the cat is clearly suffering, whereas things are not so clear in the case of the lobster. What Gunkel is cautioning us against here, therefore, is how we go about inducing from exterior resemblances of "suffering" (with reference to our own case) to an interior analogy regarding a seemingly coherent conception of sentence. If, as the Organic View suggests, we ought to view biological sentience as a

necessary condition for moral patiency, then it should follow that we have a clear way to go about identifying the entities that do or do not meet the requirement.

Consider the advent of advanced neuroimaging technology, such as functional magnetic resonance imaging (fMRI), which allows us to detect brain activity associated with blood flow. This type of technology allows us peer into the "moving parts" in the brain which may be correlated with sentience. However, talk of internal states and the talk of how we describe, scientifically, the information that an fMRI machine represents to us are two very different language games. We therefore cannot know whether two equivalent systems—one inorganic the other organic—are phenomenally different by merely putting them through a scanner. To attempt to explain what these internal states "feel like" in terms of neurophysiology and physics would be a category mistake (Powers 2013, p. 233).

This issue precludes us from being able to use "true" sentience, as specified in the Organic View, as a qualification for moral status, whether biological or artificial. The argument I have put forward, therefore, undermines the specific notion of strictly biological sentience put forward by the Organic View. What this implies, for my purposes, is that there is now conceptual space for the notion that some future artificial system may come to have a moral stake, without necessarily being sentient in the sense specified by the Organic View. We simply don't know if various entities—including other people—are only apparently or "truly" sentient. Hence, we could decide to treat all apparently sentient creatures as moral patients, which implies at some point an AI may be worthy of this type of moral attribution.

Towards a coherent account of moral patiency

From the failure of the Organic View, I would like to tentatively suggest a model for future research into moral patiency, a model which does not operate on the same biases as the Organic View. This view is broadly in line with the social-relational account presented by Coeckelbergh (2010b, 2014). Coeckelbergh claims that both "direct" (such as those based on utilitarian or deontological criteria) and "indirect" (such as those based on virtue ethics) arguments for moral patiency rest on the ontological features of the entity in question, a feature which poses significant issues for both kinds of theory (Coeckelbergh 2010b). One aspect of Coeckelbergh's argument that I think could be further developed, however, is how we might come to determine the various mental states we deem important for moral consideration (Coeckelbergh 2010a; Torrance 2013). I believe that it is possible for a defender of social-relational approaches to

---

[10] Torrance does address this issue (2014) and refers to the view that I broadly defend in this paper as "social relationism" (SR). Torrance claims that SR positions do not offer us "inherently right or wrong answers" when it comes to questions of moral patiency (2014: 12). I think this a somewhat superficial reading of SR approaches, but it is beyond the scope of this paper to go into any detail in this regard, as my focus here is concerned with the specific claims made by Torrance with regards to the criteria of moral status specifically, not realism versus social relationism more generally.

draw on certain conceptual resources developed by Daniel Dennett.

Consider how we come to infer the psychological states of others on a day-to-day basis (usually without the use of advanced neuroimaging equipment): we largely use external cues in order to make plausible predications about what might be going on in their craniums. However, this type of projection is not necessarily indicative of the "real" type of phenomenal ascription required for sentience as specified above, but at the very least it provides a predictive model that we can use to infer what might be going on in other people's heads. This methodological approach, formalised by Dennett, is known as the intentional stance (1989).[11] This "intentional stance" treats the agent in question as a rational one, and then attempts to figure out which beliefs and desires the agent ought to have in light of this capacity (ibid., p. 17). Imbued in Dennett's exposition of the intentional stance is a willingness to let go of certain outdated conceptual categories. He is willing to acknowledge that mental postulates such as "beliefs", "desires", etc. are useful for predicting behaviour, but are not good guides as to what is *really going on in the brain*. They are therefore not good theoretical entities, which is why the intentional stance must remain (and is) *non-committal* (or theory neutral) with regards to the internal structures that underlie the specific competencies that an investigator is explaining (Stich 1981, p. 44; Yu and Fuller 1986, p. 454; Dennett 2009, p. 10). In this way the intentional stance is neutral on what ontological properties need to be present in the entity under investigation – so long as we can make reasonable predications regarding the entity's behaviour by *ascribing* beliefs and desires to it, we would be correct in considering it an intentional system (Slors 1996, p. 94).

Likewise, we might be able to use the intentional stance to try to determine whether an entity in question is indeed worthy of moral consideration, based on certain behavioural cues.[12] This approach should not, however, be seen as exhaustive: it is only a helpful heuristic as to whether an entity is in fact sentient. While relying *only* on behaviouristic cues would mean that we would accord a moral stake to anything capable of, for example, mimicking pain, this would

be a mischaracterisation of my proposed usage of Dennett's methodology. My suggestion is simply that we take these behavioural cues seriously, and use them in conjunction with other relevant social-relational criteria, as opposed to only relying on the presumed capacity to have "real" qualitative states of experience or having a particular causal history, criteria that play key roles in the Organic View. In addition to behavioural cues, we might look to other cues indicative of an entity's internal constitution and what this tells us about the likelihood of this entity having the capacity for affect. This type of naturalistic approach is exemplified in the example of fish cognition above, in which our concepts and their associated usage are consistent with and do not contradict our best science (Ritchie 2008; Brown 2015). We might find that we over-ascribe the capacity for affect on this approach, but it is surely better to err on the side of caution when it comes to moral concern. The further value of creating a space in our moral and conceptual landscape for AAs is that by doing so we can perhaps solve so-called "responsibility-" and "retribution-gaps" (see Champagne and Tonkens 2013; Müller 2014; Gunkel 2017; Nyholm 2017). The former refers to cases in which it is unclear whether a human being or an AA was responsible for a moral action. The latter refers to cases in which AAs are involved in producing moral harms. In such scenarios people may feel a strong urge to punish somebody for the moral harm, but there may be no appropriate target for this punishment (Nyholm 2017).

Two more behaviouristic and functional approaches to moral ascription are the Moral Turing Test (MTT) (Gerdes and Øhrstrøm 2015) and Turing Triage Test (TTT) (Sparrow 2004). The first of these tests asks whether an artificial system "acts at least according to the ethical standards that are normally considered acceptable in human society" (Gerdes and Øhrstrøm 2015, p. 99).[13] If the system can pass such a test, then it can be worthy of moral consideration.[14] The TTT test proposes that in a "triage"[15] situation if one human person is replaced with an AA, and the moral character of the dilemma remains intact, then the AA would have achieved moral standing comparable to that of human beings (Sparrow 2004, p. 203). Both of the aforementioned propose novel ways in which we might come to understand the moral

---

[11] My decision to make use of the intentional stance is far from uncontroversial. Dennett believes that a third-person, materialistic starting point is the most appropriate one for further investigations into mentalistic concepts. This, however, can be contested on various grounds. See, for example, Nagel (1986), Ratcliffe (2001) and Slors (1996, 2015) for various philosophical issues with Dennett's account. It is far beyond the scope of the present paper to resolve these and other problems with Dennett's theory. For my purposes, however, what matters is that social-relational accounts can be amended with a theory which accounts for mental states, the details of which would still need to be worked out.

[12] These could be signs that are indicative of suffering, for example vocalizations (sighing or moaning), facial expressions (grimacing, frowning, rapid blinking, etc.) or bodily movement (being hunched over, exterior rigidity, etc.).

[13] For a critique of the Moral Turing Test, see Arnold and Scheutz (2016).

[14] Also see Wallach and Allen (2009: 70) for an exposition of the comparative Moral Turing Test (cMMT), which asks "which of these agents is less moral than the other?", as opposed to the question of which entity is the artificial agent, posed in the MTT.

[15] A situation in which a choice must be made as to which of two human lives to save.

contours of our relationships with intelligent machines in the future.

Recommendations for future research

A key issue faced by any account of moral patiency, however, is how such frameworks ought to deal with cases where the AA in question does not necessarily have humanoid features but nonetheless exhibits certain external cues that lead us to believe that it should be accorded some kind of moral concern.[16] In such cases, it is surely better to erroneously accord moral concern than to unjustifiably deny it (Wareham 2011, p. 39). A further question concerns just what exactly "machine consciousness" entails, as it need not necessarily be anything like human consciousness, making the solution to the question of machine moral patiency even more seemingly intractable. Good attempts at a philosophically coherent account of machine moral patiency are provided by Sparrow (2004), Wareham (2011), Coeckelbergh (2014), and Danaher (2017b).

In this paper I have problematised the specific sense of sentience proposed by the Organic View. To make this case I provided an exposition of the claims argued for by the Organic View, and then provided two critiques, one conceptual and the other epistemic, which served the purpose of illuminating the need for a social-relational philosophical methodology when it comes to machine moral patiency (Coeckelbergh 2010b). This new approach was introduced through the lens of Daniel Dennett's intentional stance, which could in future serve as a more philosophically coherent framework for these kinds of issues. What this implies for future research into moral patiency is that we should be careful in how we operationalise certain key concepts, such as sentience, and guard against anthropocentric fallacies as best we can. Shifting towards a social-relational methodological framework that places more emphasis on external cues might be one such way to mitigate this risk.

Acknowledgements I would like to thank my supervisor and mentor Tanya de Villiers-Botha for her insightful comments and guidance. I am also indebted to Deryck Hougaard and Lize Alberts who read earlier drafts of this paper and provided very useful feedback.

---

[16] Another arena requiring further research is the use and distribution of "entertainment" robots (Royakkers and van Est, 2015). More specifically, sex robots, which raise questions concerning the role of consent and ownership, and how (if it all) these concepts refer in this case. If we concede that such robots are AAs, can they give meaningful consent? Moreover, can we legitimately speak of acts such as "robotic rape", and punish those performing such acts (see Danaher, 2017a)? More work needs to be done at both the philosophical and regulatory levels to unpack solutions to these and other questions.

Reference

Arnold, T., & Scheutz, M. (2016). 'Against the moral Turing test: accountable design and the moral reasoning of autonomous systems. *Ethics and Information Technology, 18*(2), 103–115. https://doi.org/10.1007/s10676-016-9389-x.

Bostrom, N., & Yudkowsky, E. (2011). The ethics of artificial intelligence. In K. Frankish (Ed.), *The Cambridge handbook of artificial intelligence*. Cambridge: Cambridge University Press.

Brown, C. (2015). Fish intelligence, sentience and ethics. *Animal Cognition, 18*(1), 1–17. https://doi.org/10.1007/s10071-014-0761-0.

Chalmers, D. J. (1996). *The conscious mind*. Oxford: Oxford University Press.

Champagne, M., & Tonkens, R. (2013). Bridging the responsibility gap. *Philosophy and Technology, 28*(1), 125–137.

Coeckelbergh, M. (2010a). Moral appearances: Emotions, robots, and human morality. *Ethics and Information Technology, 12*(3), 235–241. https://doi.org/10.1007/s10676-010-9221-y.

Coeckelbergh, M. (2010b). Robot rights? Towards a social-relational justification of moral consideration. *Ethics and Information Technology, 12*(3), 209–221. https://doi.org/10.1007/s10676-010-9235-5.

Coeckelbergh, M. (2014). The moral standing of machines: Towards a relational and non-cartesian moral hermeneutics. *Philosophy & Technology, 27*, 61–77. https://doi.org/10.1007/s13347-013-0133-8.

Cohen, M. A., & Dennett, D. C. (2011). Consciousness cannot be separated from function. *Trends in Cognitive Sciences, 15*(8), 358–364. https://doi.org/10.1016/j.tics.2011.06.008.

Danaher, J. (2017a). Robotic rape and robotic child sexual abuse: Should they be criminalised? *Criminal Law and Philosophy, 11*(1), 71–95. https://doi.org/10.1007/s11572-014-9362-x.

Danaher, J. (2017b). The rise of the robots and the crisis of moral patiency. *AI and Society*. https://doi.org/10.1007/s00146-017-0773-9.

Dennett, D. (2009). Intentional systems theory. In *The Oxford handbook of philosophy of mind*, (Dennett) (pp. 1–22). https://doi.org/10.1093/oxfordhb/9780199262618.003.0020.

Dennett, D. C. (1989). *The intentional stance*. Cambridge, Massachusetts: MIT Press. https://doi.org/10.1017/S0140525X00058611.

Dennett, D. C. (1996). *Kinds of minds: Toward an understanding of consciousness*. New York: Basic Books.

Floridi, L. (1999). Information ethics: On the philosophical foundation of computer ethics. *Ethics and Information Technology, 1*, 37–56. https://doi.org/10.1023/A:1010018611096.

Floridi, L., & Sanders, J. W. (2004). On the morality of artificial agents. *Minds and Machine, 14*, 349–379. https://doi.org/10.2139/ssrn.1124296.

Gerdes, A., & Øhrstrøm, P. (2015). Issues in robot ethics seen through the lens of a moral turing test. *Journal of Information, Communication and Ethics in Society, 13*(2), 98–109. https://doi.org/10.1108/JICES-09-2014-0038.

Gunkel, D. J. (2012). *The machine question*. London: MIT Press.

Gunkel, D. J. (2017). Mind the gap: responsible robotics and the problem of responsibility. *Ethics and Information Technology*. https://doi.org/10.1007/s10676-017-9428-2.

Himma, K. E. (2009). Artificial agency, consciousness, and the criteria for moral agency: What properties must an artificial agent have to be a moral agent? *Ethics and Information Technology, 11*(1), 19–29. https://doi.org/10.1007/s10676-008-9167-5.

Johansson, L. (2010). The functional morality of robots. *International Journal of Technoethics, 1*(4), 65–73.

Johnson, D. G. (2015). Technology with no human responsibility? *Journal of Business Ethics*. https://doi.org/10.1007/s10551-014-2180-1.

Johnson, D. G., & Miller, K. W. (2008). Un-making artificial moral agents. *Ethics and Information Technology, 10*(2–3), 123–133. https://doi.org/10.1007/s10676-008-9174-6.

Johnson, D. G., & Noorman, M. (2014). Artefactual agency and artefactual moral agency. In P. Kroes & P.-P. Verbeek (Eds.), *The moral status of technical artefacts* (pp. 143–158). New York: Springer.

Leopold, A. (1948) A land ethic. In *A sand county almanac with essays on conservation from Round River*. New York: Oxford University Press.

Müller, V. C. (2014). Autonomous killer robots are probably good news. *Frontiers in Artificial Intelligence and Applications, 273*, 297–305. https://doi.org/10.3233/978-1-61499-480-0-297.

Naess, A. (1973). The shallow and the deep long-range ecology movements. *Inquiry, 16*, 95–100.

Nagel, T. (1986). *The view from nowhere*. New York: Oxford University Press. https://doi.org/10.2307/2108026.

Nyholm, S. (2017). Attributing agency to automated systems: Reflections on human-robot collaborations and responsibility-loci'. *Science and Engineering Ethics*. https://doi.org/10.1007/s11948-017-9943-x.

Powers, T. M. (2013). On the moral agency of computers. *Topoi, 32*(2), 227–236. https://doi.org/10.1007/s11245-012-9149-4.

Ratcliffe, M. (2001). A kantian stance on the intentional stance. *Biology and Philosophy, 16*(1), 29–52. https://doi.org/10.1023/A:10067 10821443.

Ritchie, J. (2008). *Understanding naturalism*. Stocksfield: Acumen.

Royakkers, L., & van Est, R. (2015). A literature review on new robotics: Automation from love to war. *International Journal of Social Robotics., 7*(5), 549–570. https://doi.org/10.1007/s12369-015-0295-x.

Singer, P. (1975). *Animal liberation: A new ethics for our treatment of animals*. New York: New York Review of Books.

Singer, P. (2011). *The expanding circle: Ethics, evolution and moral progress*. New Jersey: Princetown University Press.

Slors, M. (1996). Why Dennett cannot explain what it is to adopt the intentional stance. *The Philosophical Quarterly, 46*(182), 93–98.

Slors, M. (2015). Two improvements to the intentional stance theory: Hutto and Satne on naturalizing content. *Philosophia (United States), 43*(3), 579–591. https://doi.org/10.1007/s11140-015-9627-1.

Sparrow, R. (2004). The Turing triage test. *Ethics and Information Technology, 6*(4), 203–213. https://doi.org/10.1007/s10676-004-6491-2.

Sparrow, R. (2007). Killer robots. *Journal of Applied Philosophy, 24*(1), 62–78. https://doi.org/10.1111/j.1468-5930.2007.00346.x.

Stich, S. P. (1981). Dennett on intentional systems. *Functionalism and the Philosophy of Mind, 12*(1), 39–62.

Sullins, J. P. (2011). When is a robot a moral agent? *Machine Ethics, 6*(2001), 151–161. https://doi.org/10.1017/CBO9780511978036.021.

Torrance, S. (2007). Two conceptions of machine phenomenality. *Journal of Consciousness Studies, 14*(7), 154–166.

Torrance, S. (2008). Ethics and consciousness in artificial agents. *AI and Society, 22*(4), 495–521. https://doi.org/10.1007/s00146-007-0091-8.

Torrance, S. (2013). Artificial agents and the expanding ethical circle. *AI and Society, 28*(4), 399–414. https://doi.org/10.1007/s00146-012-0422-2.

Torrance, S. (2014). Artificial consciousness and artificial ethics: Between realism and social relationism. *Philosophy and Technology, 27*(1), 9–29. https://doi.org/10.1007/s13347-013-0136-5.

Wallach, W., & Allen, C. (2009). *Moral machines*. New York: Oxford University Press.

Wareham, C. (2011). On the moral equality of artificial agents. *International Journal of Technoethics, 2*(1), 35–42. https://doi.org/10.4018/jte.2011010103.

Yu, P., & Fuller, G. (1986). A critique of Dennett. *Synthese, 66*(3), 453–476.

Publisher's Note Springer Nature remains neutral with regard to jurisdictional claims in published maps and institutional affiliations.

Ethics and Information Technology (2021) 23:157–163
https://doi.org/10.1007/s10676-020-09535-1

ORIGINAL PAPER

The possibility of deliberate norm-adherence in AI

Danielle Swanepoel[1,2]

Published online: 13 May 2020
© Springer Nature B.V. 2020

Abstract

Moral agency status is often given to those individuals or entities which act intentionally within a society or environment. In the past, moral agency has primarily been focused on human beings and some higher-order animals. However, with the fast-paced advancements made in artificial intelligence (AI), we are now quickly approaching the point where we need to ask an important question: *should we grant moral agency status to AI*? To answer this question, we need to determine the moral agency status of these entities in society. In this paper I argue that to grant moral agency status to an entity, deliberate norm-adherence must be possible (at a minimum). In this paper I argue that, under the current status quo, AI systems are unable to meet this criterion. The novel contribution this paper makes to the field of machine ethics is first, to provide at least two criteria with which we can determine moral agency status. We do this by determining the possibility of deliberate norm-adherence through examining the possibility of deliberate norm-violation. Second, to show that establishing moral agency in AI suffer the same pitfalls as establishing moral agency in constitutive accounts of agency.

Keywords Artificial intelligence · Moral agency · Norm-violation · Norm-adherence · Constitutivism

Introduction

Often, we consider ethical behaviour as that which aligns with certain ethical norms and principles. For this to happen successfully, a person is required to meet certain conditions. First, a more obvious requirement might be that an individual should have the capacity to produce behaviour which is aligned with norms—this might not be possible if a person is either physically, mentally, or emotionally unable to—this calls into question her moral accountability. There are less obvious conditions, such as if an individual is unable to even identify norms and principles, then it seems more difficult for us to hold these individuals morally accountable since they would be violating those norms unwittingly. In addition, it might even be said that an individual is able to identify norms, but does not accept the norms as those which apply to her—this could also impact whether an act which *looks*

like norm-adherence is *actual deliberate* norm-adherence (See Kant 1785).

Finally, and most importantly, it would be difficult to hold an individual morally accountable if she *could not do otherwise*. What this means is a matter of debate (Huffer 2007; McKenna and Coates 2018; Warfield 2000). But I want to focus on one reading: for deliberate norm-adherence to take place, it must be, at a minimum, possible for the agent to endorse a norm, and have the capacity to deliberately *violate* the norm in question. When it comes to holding someone morally accountable with regards to ethical norms, we need to determine if deliberate norm-adherence is indeed possible. In this paper I argue that, under the current status quo, AI systems are unable to meet these criteria and can thus not be said to deliberately adhere to norms. This has some serious consequences, especially regarding developing AI systems as moral agents and in particular granting these entities moral accountability in society.

The novel contribution this paper makes to the field of machine ethics is first, to provide at least two criteria to showcase deliberate norm-adherence: the possibility of full norm-endorsement, and the capacity of deliberate norm-violation. Second, to show that establishing moral agency in AI suffer the same pitfalls as establishing moral agency in constitutive accounts of agency.

✉ Danielle Swanepoel
Dswanepoel@solbridge.ac.kr; dmswanepoel@zoho.com

1 University of Johannesburg, Cnr Kingsway and University Road, Auckland Park, Johannesburg, South Africa

2 SolBridge International School of Business, 127 Uam-Ro, Samseong-dong, Dong-gu, Daejeon, South Korea

Chapter 7 was originally published as Swanepoel, D. Ethics and Information Technology (2021) 23: 157–163. https://doi.org/10.1007/s10676-020-09535-1.

This paper is comprised of five sections. In "Moral consideration in AI" section, I briefly outline what moral accountability means in AI. In "What is Deliberate Norm-adherence?" section, I show how norms and principles inform actions and that there are important distinctions at play: namely, the distinction between norm-compliance and norm-endorsement. I argue that one must have the capacity for both norm-endorsement and norm-violation in order to determine if deliberate norm-adherence takes place. In "The constitutive link" section, I show that constitutive accounts of agency struggle to account for norm-violation—for similar reasons that an account of AI moral agency might fail. In "Conclusion" section, I conclude.

Moral consideration in AI

Floridi and Sanders defend the following thesis:

[T]hat AAs [artificial agents] are legitimate sources of im/moral actions, hence that A[gency] should be extended so as to include AAs, that the ethical discourse should include the analysis of their morality and, finally, that this analysis is essential in order to understand a range of new moral problems not only in Computer Ethics but also in ethics in general, especially in the case of distributed morality. (Floridi and Sanders 2004: 349)

They set out three criteria that need to be met in order for something to be considered a moral agent: interactivity, autonomy and adaptability "at a given LoA [level of abstraction]".[1] Furthermore, that in any given moral situation, one is likely to find a moral patient and a moral agent. Moral agents are those entities which qualify as being the source of moral action and moral patients are those which are recipients of moral actions—where "all entities that qualify as moral agents also qualify as moral patients but not vice versa" (Floridi and Sanders 2004, p. 350). An example of this are animals who are moral patients, but who are not moral agents.[2] Floridi and Sanders explore the possibility of AI systems holding a moral-agency status and conclude that they do, insofar as they meet the interactivity, autonomy, and adaptability criteria at a given LoA.

I disagree with Floridi and Sanders criteria upon which we are to determine moral-accountability—in particular

with their third criteria. I believe, as Coeckelbergh aptly puts it, Floridi and Sanders have lowered "the threshold for moral agency" (Coeckelbergh 2009, p. 181). Acting adaptably, according to Floridi and Sanders, is not to follow a set of predetermined orders or rules and to have the ability to change its heuristics—which, they argue, AI can do. True adaptability, I believe, presupposes, at a minimum, the ability to do *otherwise*. What this means is that I know a particular norm as one that applies to me and I have two choices in front of me: either I adhere to the norm or I reject the norm. I can freely choose either. I do not believe AI is currently in this position—even if it does have the ability to change its heuristics.

Perhaps we should view moral agency differently. Moor (2011) distinguishes between implicit and explicit ethical agency in machines. Implicit agents are those where

the machine acts ethically because its internal functions implicitly promote ethical behavior– or at least avoid unethical behavior. Ethical behavior is the machine's nature. It has, to a limited extent, virtues (Moor 2011, p. 16).

Conversely, explicit ethical agents

would be able to make plausible ethical judgments and justify them. An explicit ethical agent that was autonomous in that it could handle real-life situations involving an unpredictable sequence of events (Moor 2011, p. 17).

Under both paradigms however, the machine is programmed to behave a particular way.[3] The problem is two-fold then. First, it is impossible for AI systems to derive normative authority from facts such as rules and principles embedded in its programming language (pertaining to implicit ethical agents). But second, it is possible that AI systems, especially more advanced ones, are able to derive further principles from the original principles (as can be seen in explicit ethical agents), but it is doubtful that these are normative. Moor argues that AI may miss the mark of reaching full moral agency, as things stand, but that this should not limit us from still treating AI as ethical agents, if only in the more limited sense as implicit ethical agents (Moor 2011, p. 18), or as moral patients.

Perhaps a core problem lies in the way moral agency is being discussed pertaining to machine ethics. On the one hand, it appears that theorists such as Floridi and Sanders (2004), Gunkel (2012) have missed the mark in setting out

---

[1] The level of abstraction is determined, according to Floridi and Sanders, "by the way in which one chooses to describe, analyse and discuss a system and its context. LoA is formalised in the concept of 'interface', which consists of a set of features, the observables. Agenthood, and in particular moral agenthood, depends on a LoA".

[2] for more on moral patiency and AI, see Gunkel (2012).

[3] 1 I think it should go without saying that as AI progresses in the future, what is discussed here may no longer be relevant in the next few decades. But this is one of the unfortunate consequences of doing research in such a fast-paced industry.

---

satisfactory criteria for what counts as a moral agent—both in the human and in the machine-setting. Where Floridi and Sanders have lowered the bar regarding moral agency just enough to include AI, Gunkel has been unclear as to what counts as a moral agent but focuses instead on the divide between moral patiency and agency. On the other hand, Coeckelbergh sets out more intuitive criteria for moral agency: one should "have the capacity to freely choose one's acts […and…] one has to know the difference between right and wrong" (Coeckelbergh 2009, p. 182).

But, and more importantly, Coeckelbergh takes note of the anthropocentric practice of viewing morality as that which is exclusively attributed to human beings. When we talk about moral agents, we talk about beings that are conscious, reflective, and have capacities of rationality and free will, etc. These things are difficult to pin down in human beings and I imagine more difficult to pin down in machines. These are not pragmatic criteria upon which we can evaluate the moral status of entities. I do not pretend to have this list. What I do want to put forward is simple criteria[4] upon which we can partially determine the morality-status of an entity. I want to suggest that, in order to count as a moral agent, you need to, at an absolute minimum, be able to fully endorse a norm, and be able to *deliberately* violate a norm. Therefore, AI must, at a minimum, be able to perform actions which *go against* its programming for it to earn explicit ethical agency status—or, in this sense, moral agency.

But, if AI systems are those things which use decision-theory (or any other form of decision-making programming), especially with the aim of maximising expected utility (Hansson 1994), in their decision-making processes—then we can say that AI systems comply with norms, and in this case, norms prescribed by decision-theory. If making decisions according to a set of rules and principles is just something AI systems do, in virtue of being AI systems, then we can establish a constitutive link between the two (I discuss this constitutive link in more detail in "The constitutive link" section). Such that, if a machine did not comply with rules and principles, it would not perform actions or make decisions. Therefore, AI systems *comply* with rules and principles in virtue of *being* an AI system. If this is so, is deliberate norm-adherence possible in AI?

What is deliberate norm-adherence?

Kant explains that there is a difference between acting out of duty and acting in accordance with duty (Kant 1785). Acting out of duty is to act out of reverence for the norm. Because

not harming an individual is an ethical norm, we adhere to it out of respect of it being a norm (norm-endorsement). Acting in accordance with duty is different in the sense that it is not out of the reverence for the norm that we act but rather another inclination which is unrelated but somehow coincides with the norm—for example, say I pause at a traffic light while the light is red because I stop to answer a text on my phone. I had no intention of stopping otherwise.

Here, I want to discuss two distinctions: first, the distinction between norm-compliance and norm-endorsement; second, the distinction between deliberate norm-adherence and deliberate norm-violation.

The distinction between norm-compliance and norm-endorsement

In Thailand, the King's image is printed on the currency, and for this reason, it is illegal to step on money. If you mistakenly drop a coin, you should let it roll until it stops naturally. This is considered a norm particular to Thailand and does not apply to many other countries, say South Africa. As a South African, if I were to go to Thailand, I might find it rude that nobody bothered to step on my coin to stop it from rolling into the gutter. Others might consider it rude that I had the expectation that they *should* step on the coin to stop it from rolling into the gutter.

Citizens of Thailand understand that stepping on money is just something they should not do. Perhaps they realised this through observing the fact that others never step on money or they were disciplined from parents the first time they ever tried to do such a thing. Or, and this is more likely the case, it's just something they don't do in virtue of being born and raised in Thailand. In the first instance, when a child gets disciplined and sternly told that he should not stand on money, he understands that this is wrong and endorses this norm as one he will adhere to in the future. However, in the second case, complying with a norm in virtue of being Thai means that no actual endorsement need take place. In other words, I can comply with a norm without deliberately endorsing it as something I should be doing, it is just something I find myself doing.

The distinction between norm-compliance and norm-endorsement is straightforward.[5] Norm-compliance refers to Kant's notion of acting in accordance with the norm, and norm-endorsement (necessary for deliberate norm-adherence) is to act from that norm through embracing it as one's own. For example, if I have never been unfaithful to my partner, it could be for two possible reasons (among many): first, because I believe that cheating on my partner is wrong and

---

[4] No doubt, far more criteria need to be added in order for us to truly determine the morality-status of an entity. But here, I only want to introduce two.

[5] Many thanks to Christoph Hanisch who proposed that I use the terms norm-compliance and norm-endorsement.

this is a norm I fully endorse (a case of norm-endorsement), or; second, I don't endorse this norm, I don't believe it pertains to me, I just have not had the opportunity to cheat yet (a case of norm-compliance). Under both scenarios, I have not cheated on my partner, but for very different reasons.

Just so, when an AI system responds to stimulus and acts in accordance with the norm, we can say it is norm-compliancy. For example, if an autonomous tank has been programmed to target an enemy, then this is what it will do and if done successfully, can be said to comply with the norms of warfare. However, there is no norm-endorsement taking place. The autonomous weapon cannot defend its choice to comply with a particular norm as it cannot, conceptually, compute the meaning of the norm, nor can it compute the norm as pertaining to itself, in order for full endorsement to take place. Be this as it may, the second concern is that it cannot deliberately violate a norm—which I put forward as a partial minimum requirement for moral agency.[6]

The distinction between deliberate norm-adherence vs deliberate norm-violation

Deliberate norm-adherence has two features: first, endorsing a norm by acting from that norm, and second, to have the capacity to deliberately violate a norm. As a human moral agent, I don't harm others as I endorse this norm as one which I must respect, here, I am acting out of duty—in this sense—agential moral duty. Think, for a moment, of a person who continuously and consistently does not harm others. They do this in virtue of being moral agents—but how do we know if these acts are ones of mere norm-compliance or deliberate norm-adherence? A way to determine this is to identify if the moral agent could, if they so wished, do otherwise. For example, if I were in Thailand, maybe I would very much like to step on the coin and violate the Thai-norm, but I find myself unable to do so because I am chained to a lamppost and cannot move. In this instance, I endorse the norm as one I should adhere to, but one I would very much like to violate. But I am unable to violate. Therefore, the act of me 'abstaining' from stepping on the coin is not deliberate norm-adherence because it misses the mark of 'having the capacity to deliberately violate a norm'.

Note a distinction here. An objection to this could be that a norm may not matter to me, but I do deliberately violate that norm anyways. For instance, a sociopath doesn't care

for the norm to not hurt others, but still deliberately goes out to hurt others. Two scenarios here:

First, the sociopath understands that this is a norm for others but does not endorse the norm for himself. He cares very little about it. He still proceeds to hurt others and by doing so is violating the norm of not harming others. Second, the sociopath does endorse the norm as one that should apply to him but considers it to be one that he actively does not care about. He proceeds to purposefully violate the norm. Note, that by accepting that this norm should apply to him and that, given different circumstances—such that if he were not a sociopath, he would have deliberately adhered to this norm—is enough of an endorsement of the norm to constitute deliberate norm-violation as I mean it here.

In the first instance, the sociopath has heard others speak of the norm that we should not harm others. Perhaps he understands how this can be important for someone else. But he actively does not buy into the idea that this norm is something he should care about. Perhaps his own satisfaction and happiness takes precedence over all others. He seeks out his victims thinking that the norm of not harming people apply to others, but not to him. Maybe he even rests assured that *no one will harm him* because of this norm. And so, even if the norm-violation looks deliberate (and it is, to a large degree) it lacks the norm-endorsement we are looking for in deliberate norm-violation.

According to Johnson and Hathcock (n.d)

> The sociopath often knows exactly which kinds of action are prescribed or proscribed by morality; he or she just doesn't care. Sociopaths may possess the cognitive foundation for reliably moral behavior, but they lack the affective foundation—more specifically in this case, a motivating feeling (Johnson and Hathcock n.d, p. 55).

Johnson and Hathcock (n.d) note that moral behaviour is dependent on two factors: first, moral knowledge which requires, at a minimum, that a person is able to identify a norm and know what kind of effects acting out this norm will have. Second, and more importantly, a person must be "sufficiently motivated that they perform a moral action" (Johnson and Hathcock n.d, p. 55). In the first instance then, the sociopath has the necessary knowledge (the cognitive foundation) but lacks the motivational drive (affective foundation). It would appear that both are needed for deliberate norm-violation.

In the second instance, the sociopath has heard others speak of the norm that we should not harm others. He actively buys into the idea that this is something he should care about—thus endorsing this norm as one which should apply to him. Perhaps he recognizes himself as an active member of society and the community and because of this, accepts that he should adhere to the norm of not harming

---

[6] I appreciate that I am setting a tacit counterfactual condition here that should, in a longer account, be a) more carefully worked out, and b) related to standard versions of the Principle of Alternate Possibilities (starting with Frankfurt 1969). But I hope that the intuitive point I am trying to make here is clear without getting bogged down in the intricacies of either the vast literature on counterfactuals or on PAP.

others. But this does not automatically translate into him adhering to this norm. He accepts that this norm should apply to him but does not accept it as a norm he will adhere to at this time. This is different from the first instance in an important way. Here, the sociopath has both the cognitive foundation and the affective foundation in place. This is an instance of deliberate norm-violation.

What does this sociopath example have to do with AI? Clearly, it is problematic to label a sociopath as a fully-fledged moral agent when he may not recognise certain moral norms as those which should apply to him nor does he experience the affective foundation necessary for full-blooded moral action. Even though

> being a person is supposed to be what makes an entity a responsible agent, someone who can have duties and be the object of ethical concerns—such personhood is typically a deep notion associated with free will (Muller 2019).

It is difficult to compare the potential moral agency of an AI with the sociopath—the sociopath is already one step ahead by 'being a person'. However, the sociopath is also already failing at moral agency in two ways: either by not identifying the norms as those which apply to him, or by not being motivated to endorse those norms as his own. At best, AI could satisfy the cognitive foundation criterion insofar as AI behaves in accordance with norms. It is unclear how it would meet the affective foundation criterion. As things stand, it appears then that the best kind of ethical agency we can expect from AI is one which complies with norms in virtue of it being programmed to do so. But not much else. Until we can establish that AI *can* break the rules, so to speak, and can fully endorse norms, we are dealing with a system which is, at best, complying with norms.

The constitutive link

If the features of deliberate norm-adherence is norm-endorsement and the capacity to deliberately violate a norm, then, surprisingly, constitutive accounts of agency are also unable to account for deliberate norm-adherence. The very pitfalls these constitutive accounts of agency encounter are similar pitfalls we encounter when we try to establish the moral-agency status of AI systems.

According to some constitutive accounts (Korsgaard 2009; Railton 2003; Velleman 1996), adhering to norms is just something an agent does in *virtue* of being an agent. Such that, insofar as you are an agent, you adhere to norms. Such accounts, 'agent- constitutive accounts'[7] try to argue that normative standards, aims or motives are constitutive features of agency. Furthermore, that these features are the source of the normative grip or of the authority norms (such

as moral norms) have over our beliefs, desires, deliberations, motivations and actions (Bratman 2007; Davidson 1963; Railton 2003; Velleman 1996, 2004). The outcome is that if we were not to possess these specific features of agency, we would not be agents (Ferrero 2009). Consequently, if the norms, or standards, or aims didn't have the requisite grip on us, our agency is likewise undermined.

The problem with constitutivist accounts, however, is that they struggle to explain deliberate norm-violation. For instance, In Korsgaard's account, if an agent fails to adhere to norms, she essentially fails at the game of agency (Korsgaard 2009). Arguably, if an account of agency is unable to explain deliberate norm-violation, then it cannot be said to explain deliberate norm-adherence. And so, if constitutive accounts run the argument that you adhere to norms (and not violate norms) in virtue of being an agent, then there are at least two undesirable consequences to be had here: first, if I am right that deliberate norm-adherence presupposes the possibility of deliberate norm-violation, such a theory deprives us of the ability to deliberately adhere to norms at all. Or, worse still, when we do violate norms, we forego agency (Enoch 2006).[8]

If a constitutive account is only able to explain deliberate norm-adherence in terms of agency, then instances of norm-violation (both unintentional and deliberate) potentially spell out instances of non-agency. This is counterintuitive as there are many instances, in everyday life, where agents violate norms without undermining agency. Intuitively at least, if I killed someone, I wouldn't lose my agency. Nor do I lose it when I do something irrational—say pick up my broken phone to call the phone company to ask them to come and fix it. I say 'intuitively' deliberately. If you are a constitutivist like Korsgaard, you would perhaps think that I do lose my agency in these instances. But this would simply beg the question at hand. In fact, when I deliberately violate a norm, I am practising my agency to its fullest. As far as an individual follows norms in virtue of being an agent, any non-adherence spells out the negative consequences outlined above.

Constitutivists often, and tacitly so, view norm-adherence as encompassing either norm-compliance and deliberate

---

[7] Acknowledgement to Veli Mitova who used this term in discussions we had.

[8] There are, of course, various ways of getting around this, such as voluntarist approaches or a hybrid theory which sees the merging of voluntarist and constitutivist features (Bratman 2007; Korsgaard 2008; Katsafanas 2013; Rosati 1995, 2003, 2016; Tiffany 2012). However, these approaches face their own set of problems which I cannot go into here for brevity sake. This paper is not particularly interested in proving the legitimacy of constitutivism, but is rather only interested in the constitutive relationship between agency and norms, where norms are adhered to in virtue of them being part and parcel of the features constituting agency.

norm-adherence. This is problematic. Norm-compliance can take many forms, such as absent-mindedly crossing the road when the light is green; or not stepping on a coin in Thailand because you dislike the sound of metal hitting a pavement. But norm-compliance does not always exhibit an instance of norm-endorsement. This is not where we should be pinning agency-fulfilment. Deliberate norm-adherence is where agency is fully exercised. It looks like this: I endorse that X is a norm, I further accept that X has normative authority over my actions. Also, I know that I could violate X if I so wanted to, but I choose to adhere to X…or I choose not to. Either way, agency remains intact.

What then does this mean for moral agency status? Should we be so quick to consider someone morally praiseworthy just because they comply with norms if it is entirely possible they could not do otherwise? As things stand, Current constitutivism, at least in the strictest sense, cannot account for deliberate norm-violation, and so, arguably, cannot really account for deliberate norm-adherence.

I want to take two things from the above discussion about constitutive accounts of agency: first, the notion that adhering to norms is somehow constitutive, and second, that norm-endorsement, and the capacity to violate norms are key elements for deliberate norm-adherence. If we think about AI systems as those things which follow rules and principles in virtue of being the programmes or machines that they are, then it is possible to establish a constitutive relationship between the machine and the compliance with principles or rules in the same way we can establish a constitutive relationship between adhering to norms and agency. However, I argue, *this* constitutive relationship is one of norm-compliance, and maybe, at a stretch, norm-endorsement.

First, we perceive machines as those things which act in accordance with rules and principles, and sometimes we think that because it acts in the world, it is deserving of moral agency status. Rather, I argue that we should think of machines as complying to norms in the same way as the more traditional constitutive view of agency sees agents as individuals who comply with norms and endorse norms. But what is missing from this story is that these individuals are incapable of deliberate norm-violation and incapable of full norm-endorsement, and so: how do they deliberately perform good actions when they are incapable of performing bad ones?

To accommodate deliberate norm-violation in AI systems, at least in the proper sense (and in the way to prove deliberate norm-adherence), we need to show how norm-endorsement and norm-violation is possible in AI systems. As things stand, this doesn't seem likely. As Castelfranchi et al. (2000, p. 364) maintains, "if the conventions and norms are hard-wired into the agent's protocols it cannot decide to violate the norms". Why not? Because AI systems have no way of deriving normative authority from facts about the rules and principles it has been programmed to follow; second, AI systems do not have a grasp of the concept of normativity in order for norms to apply to it; third, AI systems do not meet the affective foundation criterion. AI systems are confined and restricted to their programming and until norm-endorsement and deliberate norm-violation are made possible in these systems, deliberate norm-adherence in these systems remains doubtful.

Conclusion

What I offer in this paper is a minimal explanation as to why we should not consider AI as having moral agency status, or at least should do so in a very limited sense. Furthermore, that we should not consider AI, as things stand, as being those entities which are capable of being held morally accountable. I argued for this in the following way. First, I provide a brief background of relevant literature regarding the moral-status of AI systems—showing how the AI systems are bound to follow rules and principles which are part and parcel of their programming. Second, I examined the distinction between norm-compliance and norm-violation, showing that it is possible that actions can appear to be those of adhering to norms but that, as long as one is unable to endorse norms and deliberately violate norms, then the norm-adherence is ingenuine (at best).

I then proceeded to tie in the notion of norm-adherence to that of constitutivism and showed that, under constitutivist accounts of agency, norm-adherence is a confirmation of agency because things adhere to norms in virtue of being agents and showed that to exercise agency to its fullest, norm-endorsement is crucial. I explored the notion of a constitutive tie between norm-compliance and AI machines (in the same way adhering to norms can be seen as constitutive of agency), but where these constitutive accounts of agency could have a way out because an agent is capable of norm- adherence in virtue of norm-endorsement, AI systems are not capable of this. Because of this inability, at best, an AI system can be said to comply with norms insofar as it acts in accordance with a norm, but in no way is it deliberate norm-adherence because it has no ability to deliberately *violate* a norm. Thus, we should abstain from perceiving AI as having moral agency status until there is a suitable change in the current status quo.

Acknowledgements Many thanks to Veli Mitova for her encouragement, invaluable feedback and assistance. Thanks to Thaddeus Metz for his advice regarding article writing and feedback on a related project which greatly informed this one. Further thanks to Samuel Segun for his very helpful comments. Finally, thanks to all at the university of Johannesburg and SolBridge International School of Business who facilitated and assisted.

References

Bratman, M. (2007). *Structures of agency*. New York: Oxford University Press.

Castelfranchi, C., Dignum, F., Jonker, C., & Treur, J. (2000). *Deliberative normative agents: Principles and architecture. Intelligent agents* (pp. 364–378). Berlin: Springer.

Coeckelbergh, M. (2009). Virtual moral agency, virtual moral responsibility: On the moral significance of the appearance, perception, and performance of artificial agents. *AI and Society, 1*, 10–25. https://doi.org/10.1007/s00146-009-0208-3.

Davidson, D. (1963). Actions, reasons, and causes. *The Journal of Philosophy, 60*(23), 685–700.

Enoch, D. (2006). Agency, Shmagency: Why normativity won't come from what is constitutive of action. *Philosophical Review, 115*(2), 31–60.

Ferrero, L. (2009). Constitutivism and the inescapability of agency. *Oxford Studies in Metaethics, IV*, 303–333.

Floridi, L., Sanders, J. W. (2004). On the morality of artificial agents. *Minds and Machines, 14*, 349–379.

Frankfurt, H. (1969). Alternative possibilities and moral responsibility. *Journal of Philosophy, 66*(23), 829–839.

Gunkel, D. (2012). *The machine question: Critical perspectives on AI, robots, and ethics*. Cambridge: MIT Press.

Hansson, S. (1994). *Decision theory: A brief introduction*. Stockholm: Royal Institute of Technology.

Huffer, B. (2007). Actions and outcomes: Two aspects of agency. *Synthese, 157*, 241–265.

Johnson, A., Hathcock, D. (n.d). Study abroad and moral development. *Ejournal of Public Affairs, 3*(3), 52–70.

Kant, I. (1785). Groundwork for the metaphysics of morals. In: A. Wood (Ed.) *Groundwork for the metaphysics of morals*. New York: Yale University.

Katsafanas, P. (2013). *Agency and the foundation of ethics: Nietzschean constitutivism*. Oxford: Oxford University Press.

Korsgaard, C. (2008). *The constitution of agency. Essays on practical reason and moral psychology*. Oxford: Oxford University Press.

Korsgaard, C. (2009). *Self-constitution: Agency, identity, and integrity*. Oxford: Oxford University Press.

McKenna, M. & Coates, J., (2018). *Compatibilism*. [Online]. Retrieved March 25, 2019, from https://plato.stanford.edu/archives/win2018/entries/compatibilism/.

Moor, J. (2011). The nature, importance, and difficulty of machine ethics. In M. Ethics (Ed.), *Anderson & Anderson* (pp. 13–20). New York: Cambridge University Press.

Muller, V. (2019). Ethics of AI and robotics. Retrieved August 15, 2019, from https://www.researchgate.net/project/Ethics-of-AI-and-Robotics-for-Stanford-Encyclopedia-of-Philosophy.

Railton, P. (2003). On the hypothetical and non-hypothetical in reasoning about belief and action. *Ethics and practical reason* (pp. 53–80). Oxford: Clarendon Press.

Rosati, C. (1995). Naturalism, normativity, and the open argument question. *Nous, 29*(1), 46–70.

Rosati, C. (2003). Agency and the open question argument. *Ethics, 113*(3), 490–527.

Rosati, C. (2016). Agents and "shmagents" an essay on agency and normativity. In R. Shafer-Landau (Ed.), *Oxford studies in metaethics 11* (pp. 182–213). Oxford: Oxford University Press.

Tiffany, E. (2012). Why be an agent? *Australasia Journal of Philosophy, 90*(2), 223–233.

Velleman, D. (1996). The possibility of practical reason. *Ethics, 106*(4), 694–726.

Velleman, D. (2004). Replies to discussion on the possibility of practical reason. *Philosophical Studies, 121*, 225–238.

Warfield, T. (2000). Causal determination and human freedom is incompatible: A new argument for incompatibilism. *Nous, 34*, 167–180.

Publisher's Note Springer Nature remains neutral with regard to jurisdictional claims in published maps and institutional affiliations.

 Springer